The Washington Iron Works of Franklin County, Virginia

Excavated near the furnace, this fragment *(top)* from a eighteenth-century stove plate with the letters *WAS* visible is the only known example of a product bearing the name of the Washington Iron Works. The second stove plate fragment *(bottom)* decorated with Pennsylvania German folk art designs is the only dated piece from the Washington Iron Works yet discovered.

The Washington Iron Works of Franklin County Virginia 1773–1850

John S. Salmon

Virginia State Library • Richmond • 1986

Library of Congress Cataloging in Publication Data
Salmon, John S., 1948–
The Washington Iron Works of Franklin County.

Bibliography: p.
Includes index.
1. Iron industry and trade—Virginia—Franklin County—History. 2. Washington Iron Works—History.
I. Title.
HD9517.V52S25 1986 338.7'6691'0975568 86–13166
ISBN 0–88490–137–8

Standard Book Number: 0–88490–137–8

Virginia State Library, Richmond, Virginia.
Printed in the United States of America.

Contents

Preface

Prior to the Civil War, Virginia was one of the leading iron-producing states in the Union, and during the eighteenth century Virginia had exported more iron than all the other British North American colonies combined. Although by 1861 the giant Tredegar Iron Works, in Richmond, was the visible symbol of the Old Dominion's iron industry, scores of small charcoal blast furnaces and forges tucked among the hills and streams of western Virginia were a continuing source of iron products as they had been for decades.[1] This book traces the history of the Washington Iron Works, in Franklin County, a small charcoal furnace typical in size and operation. The second Virginia works erected south of the James River and the first major industry of Franklin County, the Washington Iron Works differed from other antebellum Virginia ironworks in two ways: its continued vitality (in a period when iron-manufacturing companies rose and fell with alarming regularity) and the unusual talents and personalities of the men associated with it. The owners of the ironworks and their extended families were among the most influential inhabitants of the county in the antebellum period, and their role in the county's history is remembered there even today.

In Rocky Mount, Virginia, stands the house of Peter Saunders, ironmaster of the Washington Iron Works. When Ann Saunders Joplin (great-great-granddaughter of Peter Saunders's brother Samuel) found caring for the big old house too taxing, she agreed to sell it in 1971 to J. Francis Amos. The proud new owner, a friend from student days at the University of Virginia, inspired my interest in the house and the ironworks. Within a few years, my research had raised a new question for every one I answered and I set out to write a thorough history of the Washington Iron Works. This book is the result.

1. The story of the Tredegar Iron Works has been well told by Charles B. Dew in his *Ironmaker to the Confederacy: Joseph R. Anderson and the Tredegar Iron Works* (New Haven, Conn., 1966). Other major individual nineteenth-century American ironworks include the Hopewell works in Pennsylvania and the Saugus works in Massachusetts. Kathleen Bruce's *Virginia Iron Manufacture in the Slave Era* (New York, 1931) is now more than fifty years old, and Virginia's smaller ironworks have, in recent years, only been mentioned in articles about industrial slavery such as S. Sidney Bradford, "The Negro Ironworker in Ante Bellum Virginia," *The Journal of Southern History* 25 (1929): 194–206 and Charles B. Dew, "David Ross and the Oxford Iron Works: A Study of Industrial Slavery in the Early Nineteenth-Century South," *The William and Mary Quarterly*, 3d ser., 31 (1974): 189–224.

I am grateful for assistance from many quarters as I conducted my research. In Rocky Mount, the late Ann Saunders Joplin shared good-humored anecdotes about her family; Gertrude C. Mann, long the unofficial historian of Franklin County, helped me unravel the mysteries of local history; and Anne Carter Lee cheerfully debated many fine points of interpretation that my research raised. William Lee, of Bleak Hill in Franklin County, made the Saunders family papers in his possession available for study. The staffs of the Virginia Historical Society, the University of Virginia, the University of North Carolina, and Duke University assisted my research in their collections, and my colleagues in the Archives and Records Division of the Virginia State Library were helpful and patient as they listened time and again to talk of Franklin County and the ironworks. To Frank Horton, director of the Museum of Early Southern Decorative Arts, in Winston-Salem, North Carolina, I am grateful for a delightful tour of the museum, a lengthy conversation about southern iron, and access to the museum's newspaper index and photographic files. William Robinson, on whose property the Washington furnace stands, graciously permitted access to the site. My wife, Emily, assisted the later stages of research and gave support and encouragement all along the way. Francis and Laquita Amos offered endless hospitality and shared many hours riding over back roads, hiking through woods, climbing down cliffs, and crawling along riverbanks to locate the sites of now-vanished furnaces and forges. At the Virginia State Library, Jon Kukla, Rosalie A. West, and Emily J. Salmon edited the manuscript for publication, and W. Donald Rhinesmith saw the book through press. Fine people such as these eased the difficulties of my research, but any errors in this book are my responsibility, not theirs. Finally, to my parents, Henry S. Salmon and Janice Salmon, I owe a debt of gratitude for their interest in my work and their encouragement to see it through to completion.

John S. Salmon

Glossary

AIR FURNACE, a furnace in which iron is melted by a flame originating from fuel burned at one end, passing over a hearth in the middle, and exiting at the other end.

BLAST FURNACE, a furnace in which a blast of air is blown through the fuel to increase the combustion rate.

BLOOMERY, a small furnace in which iron is melted from ore and stirred by hand using metal crowbars to produce a ball, or bloom, of iron.

COLLIER, a worker who makes charcoal.

FOUNDER, an ironworker who makes castings.

GUDGEON, a metal pin on which a wheel turns at the end of a wooden axle.

HOLLOW WARE, cast iron that is hollow in form, such as pots, skillets, etc.

PIG IRON, thick bars of iron formed when molten iron flows from the furnace into long channels in the casting bed, called sows, and into shorter side channels, called pigs.

PUDDLING, a process for producing wrought iron by mechanical agitation of a bath of molten pig iron with iron oxide to reduce the impurities in the pig iron.

REFINERY FORGE, a building with fireplaces in which pig iron is reheated and pounded with hammers to remove impurities.

RUN-OUT, a fireplace in a refinery forge in which pig iron is reheated until it becomes molten again.

The Washington Iron Works of Franklin County, Virginia

1

Iron Manufacturing in Colonial Virginia

Virginia iron manufacturing began almost as soon as the first settlers arrived. Within a year of the 1607 landing at Jamestown, iron ore was discovered and shipped to England, where it was smelted into pig iron. Encouraged by the good quality of the ore, the colonists decided to establish their own iron industry. They selected Falling Creek, sixty-five miles up the James River from Jamestown, as the site for their ironworks, and in 1619 skilled ironworkers were sent to the colony from England. That same year saw the arrival in Virginia of the first Africans, who later would play an important role in southern iron manufacture.

Owing to various delays it was not until 1622 that the first ironworks in the New World was ready to begin production. Judging from the length of time it took to build, the ironworks probably consisted of a blast furnace to melt the iron from the ore and a refinery forge to hammer out bar iron. But on 22 March 1622, before the works could begin operation, the Indians under Opechancanough attacked the settlers, destroying the ironworks and killing the workers. This sudden attack forced the colonists to retrench, and postponed for a century the successful establishment of the iron industry in Virginia.

Other colonies had less difficulty beginning their iron industries. In Massachusetts Bay ironworks were erected at Lynn and Braintree by the middle of the century, and by 1656 Plymouth Plantation had built a bloomery forge. Colonists in Connecticut built a blast furnace and forge at New Haven, which began operation in 1657, and before the century ended citizens of Rhode Island and New Jersey had also constructed ironworks.

In Virginia efforts to manufacture iron were not resumed until the first quarter of the eighteenth century, when Governor Alexander

Spotswood imported German miners and settled them at Germanna in present-day Spotsylvania County. It is not known just when Spotswood's ironworks began production, although 1716 is the traditional date. By 1732 there were four Virginia ironworks in operation, and in that year William Byrd II of Westover, thinking of building his own works, visited three of them.

Byrd stopped first at the Fredericksville works in Spotsylvania County, owned by a company of Virginians and managed by Charles Chiswell, who showed Byrd the blast furnace and mines. He then traveled to Governor Spotswood's home, where he was given a tour of the mines and furnace near Germanna. He also visited Spotswood's air furnace on the Rappahannock River, where pig iron from the blast furnace was re-melted and cast into hollow ware. Virginia's fourth furnace, which Byrd did not see, was the Accokeek Furnace owned by the Principio Company of Maryland and built on the land of Augustine Washington in Stafford County. In 1732 all of the Virginia ironworks consisted of furnaces only; no refinery forges had been constructed.

Byrd's description of the ironworks he visited might be applied to the typical Virginia iron plantation of a hundred years later, except for the absence of forges. At each he found, beside the furnace and mines, a planned community that was as self-sufficient as possible. Byrd saw heaps of iron ore and roasting charcoal, a storehouse for the charcoal, carts for transporting ore and pig iron, cornfields, and a gristmill.[1] He found that "120 slaves, including women, were necessary to carry on all the business of an iron work, [including] carters, colliers, and those that planted the corn."[2] Byrd also learned from Chiswell that

> besides the founder, the collier, and miner, who are paid in proportion to their work, the company have several other officers upon wages: a stocktaker, who weighs and measures everything, a clerk, who keeps an account of all receipts and disbursements; a smith to shoe their cattle [oxen] and keep all their iron-work in repair; a wheelwright, cartwright, carpenter, and several carters.[3]

Once Spotswood had set the example, other ironworks were erected following his model. By 1738 John Tayloe, of Richmond County, had built the Neabsco Furnace in Prince William County, and in 1755 his son John Tayloe II, in partnership with Presley Thornton and John Ballendine, opened another furnace and a forge in Prince William County on Occoquan River.

In addition to these improvements in tidewater Virginia, German and Scotch-Irish settlers, moving from Pennsylvania south through the Valley of Virginia and into the piedmont, began to establish small ironworks as early as 1742. In 1760 John Miller built a furnace on Mossy Creek in

Making charcoal in heaps.

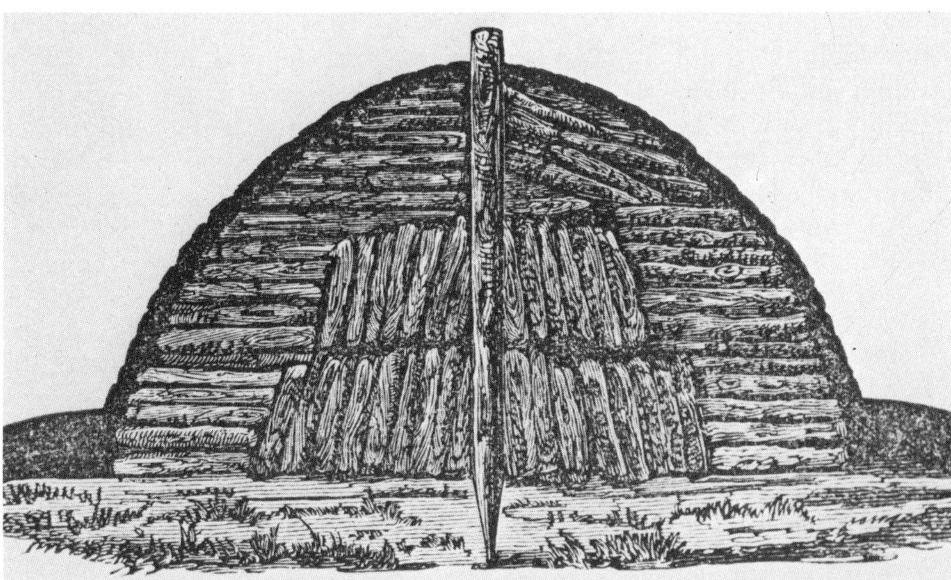

Section of a charcoal work—piling the wood.

Eighteenth-century and antebellum iron production required tons of charcoal. These diagrams from Frederick Overman's *The Manufacture of Iron* (Philadelphia, 1850) illustrate a pattern of stacking wood into heaps to be burned into charcoal—"a mode of working very much in use" that with "care and attention, furnishes good results."

Augusta County, and five years later he added a forge. Also in 1760 Isaac Zane constructed a furnace and forge in Frederick County called the Marlboro Works. Three furnaces were erected in Albemarle County in the 1770s. By the beginning of the revolutionary war, then, the Virginia iron industry was well established in the Valley, in tidewater, and in the piedmont north of James River.

What of ironworks south of James River? Here the record is vague, but it appears that a bloomery forge may have been built as early as 1768. On 21 April of that year Nicholas Davies, of Bedford County, advertised in the *Virginia Gazette*

> that he hath very lately discovered on his land, in Bedford county, four large iron banks, two of which may be more properly called iron mountains, all within three miles of James river, convenient to water carriage, and good places for a mill on the said water, and plenty of timber of all sorts; and he is determined, as soon as he possibly can, to erect such kind of iron works as he can compleat, in order to make iron, and as he is unacquainted with such affairs, is desirous of taking in some partners that are skillful in such works, and capable of assisting him in contriving and carrying on so useful an undertaking.[4]

It is not known whether Davies was successful in acquiring partners and building his ironworks, but the property was later purchased by three other partners, Robert Harper, of Berkeley County; Benjamin Elledge, of Bedford County; and Thomas James, of the Carlisle ironworks in Cumberland County, Pennsylvania. Harper and James announced in the *Virginia Gazette* on 22 October 1772 that the "Oxford Forge, or Bloomery," as well as a sawmill near the forge and 6,000 acres of land, would be sold on 1 April 1773:

> There are several banks of ore, sufficient for any number of iron works the owner or owners may think proper to erect. The quality of the iron that has already been made in the bloomery, or blooming way, is not inferior to any on the continent. There are two seats for furnaces, the farthest not exceeding a quarter of a mile from the ore banks. . . . The country in which the said works are situated is very extensive, and as there are no iron works to intercept their market, the demand for iron will be very great.[5]

A brief notice in the *Gazette* on 10 June 1773 informed the public that the Oxford Bloomery had been "made over, by Deed of Trust, to James and Robert Donald and Company [Scottish merchants], as Security for a Debt due them for Dealings at their Rocky Ridge and Bedford Stores."[6]

Hereafter the history of the ownership of the Oxford Bloomery is unclear, but by 1776 it was in the possession of David Ross, a wealthy

Petersburg merchant.[7] Regardless of the confusion over its ownership, there is no doubt that the Oxford ironworks was important for being the first of its kind south of the James River. But the iron monopoly of which Harper and James boasted in the *Gazette* was short-lived, for in 1773 a second bloomery was built about sixty miles southwest of the Oxford works by John Donelson, a frontiersman and surveyor of Pittsylvania County.

2

John Donelson and His Bloomery

John Donelson was born on 7 April 1725 in Somerset County, Maryland.[1] His father, John Donelson, Sr., had immigrated to America from England in 1716, and first had settled in Delaware, where he married Catherine Davies.[2] His son eventually moved across the Maryland-Virginia border to Accomack County and married Rachel Stockley, daughter of Alexander Stockley. Young Donelson must have found life in old Accomack too tame, for in 1744 he claimed 200 acres of land on Sandy Creek in Brunswick County,[3] and soon moved westward into the wilderness. He finally settled on Banister River in what was to become Pittsylvania County, and acquired land there. In 1761 he obtained permission from the Halifax County Court to build a gristmill on Thorn Creek.[4]

Donelson appeared before the county court of Halifax on 17 September 1761 and, after "producing a Commission from the President and Masters of the College of William and Mary appointing him Deputy Surveyor of this county,"[5] he was sworn in as deputy surveyor. Three years later, on 17 May 1764, he became surveyor for Halifax County.[6] The territory for which he was responsible was huge, including the present counties of Halifax, Pittsylvania, Henry, Patrick, and Franklin.

Frontier society was quick to recognize men of ability, and young Donelson soon rose in prominence. Besides his duties as surveyor, he had, on 17 February 1764, taken on those of a vestryman for Antrim Parish.[7] On 26 June 1767, at the first meeting of the court for newly formed Pittsylvania County, John Donelson took the oath of a justice.[8] He acquired three offices on 26 November 1767, when he was sworn in as a vestryman for Camden Parish, a colonel of the militia, and the surveyor for Pittsylvania County.[9]

Even before Donelson became a surveyor for Halifax County, however, other explorers had discovered iron ore in what is now Franklin County. On 14 December 1750 David Wilson entered a claim with the county surveyor for 400 acres of land near Jacks Mountain, "including several small Draughts of Pig River enter'd on account of a Suppos'd Mine."[10] Although this entry does not specifically mention iron, other evidence indicates that it was not long before iron was being mined.

On 3 May 1753 John Wilcox, of Pennsylvania and North Carolina, had 400 acres surveyed "on Iron Mine Branch of Pigg River,"[11] and on 26 April 1754 he had another tract of 403 acres surveyed.[12] Having thus established his claim to these tracts, Wilcox then followed the usual procedures for obtaining a patent from the colony for uninhabited land. Plats of the surveys were sent to the secretary of the colony in Williamsburg; the secretary compared the plats to others from the same area to be sure no one else was claiming the same land, and collected a small fee from Wilcox before the patents were issued. Such were the complications of life on the frontier that Wilcox did not obtain legal patents for his land until ten years later, on 27 June 1764.[13] Undoubtedly he, like other frontiersman, did not let the lack of a legal title prevent him from using the land in the meantime, as his pattern of land acquisition demonstrates. Wilcox already had patented, on 10 June 1760, 400 acres of land on lower Iron Mine Branch (now Furnace Creek) and Pigg River;[14] with his patents of 1764, he owned all of the iron ore deposits on Iron Mine Branch from Pigg River to the foot of Grassy Hill, a total of 1,203 acres.

No doubt Wilcox intended to construct an ironworks on his property, but he never carried out his plan. Born on 21 June 1728 in Chester County, Pennsylvania, Wilcox had moved to Cross Creek in Cumberland County, North Carolina, by 1759. He later moved to Deep River in Chatham County, where he built a sawmill and gristmill in 1768.[15] He became involved in the Regulator Movement that flourished in North Carolina between 1769 and 1771, and by 1772 he and Thomas Hadly owned a dry goods business in Chatham County. On 31 March 1771 he wrote from Deep River,

> I have got a Good forge in the best part of the Cuntry. The Oare I Principly depended on proovs hard to be Got and . . . about ten mils from the forks I have discover a larg Mounten of Oar . . . with Every Convaincy Nesery for a furnice.[16]

By 1776 he had built a blast furnace on his Deep River property.[17] Wilcox ultimately overreached himself and failed in his several enterprises, and his troubles may have begun with his land speculation in Virginia.

In 1768 Wilcox agreed to sell to John Donelson the two tracts of land he had patented in 1764, but the deal was never consummated, and Wilcox, who had traveled to Virginia, returned to North Carolina. Before he left, however, he "placed in the hands of Donelson as much money as would be sufficient to discharge the arrears of quitrents then due on the land, which Donaldson [*sic*] promised to apply to that purpose."[18] It was crucial that the quitrent—a token rent that had been paid annually to the lord of the manor in the Middle Ages, and that was paid to the colonial government to keep the title to a patent alive—be paid. If it were not, the title would lapse, and anyone could claim the land. When a man named John Cox filed a petition for Wilcox's land with the General Court, claiming that Wilcox had forfeited it by not paying quitrents to the Crown, Donelson wrote Wilcox on 8 August 1769 to tell him of the impending trial.[19]

Donelson did not tell Wilcox that ten days earlier he had purchased Cox's right to the land if Cox won the case. At the trial, Donelson testified that he had paid the quitrents, but his testimony was thrown out when his transaction with Cox was revealed, because this made him an interested party to the suit. Since Wilcox did not attend the trial, and since no one else testified in his behalf, the General Court in 1772 ordered his land forfeited to Cox. Cox then assigned it to Donelson, who was issued patents for both tracts on 5 July 1774.[20] Now that Donelson had secured the heart of the iron-bearing district around Furnace Creek and Pigg River, he decided to consolidate it with several adjoining tracts of land into one tract totaling 3,988 acres, and had a survey made for that purpose.[21] As far as Donelson was concerned, the episode with Wilcox was over. Sixteen years later, however, Wilcox would return, claiming fraud, and take the ironworks' new owners to trial.

Although the part of Pittsylvania County that is now Franklin County was largely wilderness in 1773, it was not altogether uninhabited. Settlers began arriving in the 1740s, and in the third quarter of the century the pace of migration quickened. Most of the early pioneers emigrated from eastern Virginia, but there were others who traveled up the Valley of Virginia on the Great Wagon Road, which led from Philadelphia to the Carolinas. Many of these German and Scotch-Irish colonists continued on their way to the Carolina backcountry, but a few left the road as it passed through present-day Franklin County, and remained to farm.

The pioneers who settled in the Virginia backcountry were limited in their possessions to what they could carry with them on horseback or in a wagon. Stores were scarce, roads were terrible, and the cost of transportation was high. A new settler was forced by necessity to rely upon himself and his neighbors to clear the land, build a dwelling, and earn a livelihood. His most valuable possessions were his tools—axes, hatchets, mattocks, froes, wedges, augers, plows, hoes, and shovels. Also high on his list of essential items were hollow-ware cooking utensils: frying pans, skillets, dutch ovens, pots, and kettles. The Germans carried with them their heavy five- and ten-plate stoves. All these things were made of iron, and their loss through damage or wear meant added hardship in an already hard life. Given the problems and costs of transportation, a local

Locations in present-day Rocky Mount, Virginia, associated with the Washington Iron Works include (1, 2, 3) the sites of ore pits, (4) the ironmaster's house, (5) the furnace, (6) the forge site, and (7) the county courthouse. Original map (1984) courtesy of the Virginia Department of Highways and Transportation.

source of replacement hollow ware and metal would have been a real boon to the pioneer, and a constant source of profit to its owner.

To answer the demand, John Donelson built a bloomery on his new land, giving the Oxford Bloomery in Bedford County a competitor. Donelson hired four men, John Hollaway, Charles Hollaway, Amos Span, and Thomas Bolton, to construct and operate his ironworks, and provided six slaves—Dick, Harry, James, Judith, Moody, and Nell—to mine and cart the ore and feed the workers.[22] In addition to the bloomery and forge, Donelson's ironworks probably had storage sheds for charcoal and extra iron ore, a small warehouse for bar iron, and dwellings for the slaves and white workmen. By 1773 a small industrial village stood where there had been only wilderness the year before.

The exact site of the bloomery has not been determined. Donelson may have built it close to the ore deposits on the upper reaches of Iron Mine Branch. On the other hand, if the bloomery was constructed of durable native granite instead of brick, perhaps it was built on the site of the present furnace on Furnace Creek. Across the creek from the furnace, on Scuffling Hill, there is an outcropping of granite—one of the few places it occurs in the county. The ore banks are close enough to the furnace site that hauling the ore to a bloomery there would have been only a minor inconvenience.

Mining the ore was not difficult. The miners chose a likely spot where loose ore was lying on the ground and then they dug a pit and picked the ore by hand. Several of these ore pits are visible today near the Shoaf Springs area of Rocky Mount off Diamond Avenue. The Shoaf Springs iron ore deposit consists of a sandstone that is highly metalliferous.[23] The ore can be seen in its natural state as a dense rust-colored rock in the southern face of the Norfolk and Western Railroad cut just west of the intersection of Diamond Avenue and the railroad tracks. There is a very large pit, mostly filled in for the construction of houses, just up the hill from the railroad cut, on the north side of Windsor Drive. Also clearly visible in the woods to the east of the railroad cut and down the slope from the tracks are several smaller pits, some of which still have chunks of ore lying near them.

Once the miners had dug the ore from the ground, workers broke it up and carried it to the bloomery, where it was smelted into blobs of semimolten iron called blooms. The amount and quality of the iron depended on the quality of the ore and the care with which the bloomery was constructed.

Building a bloomery forge, which resembled a large blacksmith's forge, was not beyond the capabilities of a skillful chimney builder, provided he was guided by someone who was acquainted with the operation. Upon a thick six-to-eight-foot-square hearth (in this case probably built of granite although other stone could be used) the mason

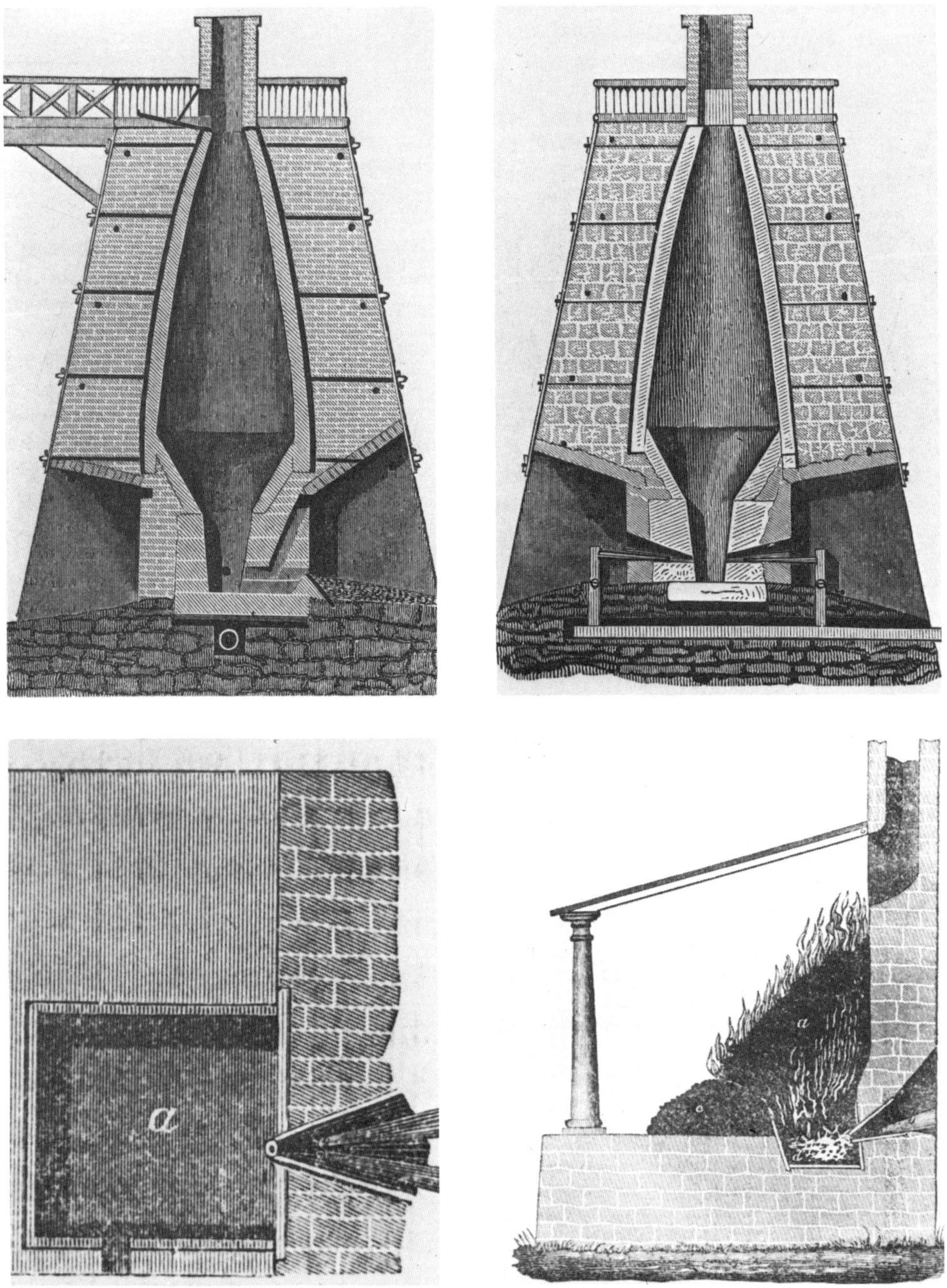

Diagrams from Frederick Overman's *The Manufacture of Iron* (Philadelphia, 1850) show *(top right)* a fore-to-aft cross section of an antebellum blast furnace with the wooden bridge at the top used to transport charcoal, iron ore, and limestone to the mouth of the furnace, the work arch at the right, and an air pipe beneath the furnace; *(top left)* a side-to-side view of the tuyere arches through which blasts of air entered the furnace; *(bottom left)* the ground plan of a forge fireplace; and *(bottom right)* a typical small bloomery furnace.

erected a fireplace and chimney in a back corner, with a roof sloping forward from the front of the chimney toward the ground to help conduct the heat of the fire into the chimney. The fireplace was about twenty-four to thirty inches square, and was recessed into the hearth to a depth of from fifteen to twenty inches. The most critical part of the bloomery was the tuyere, an opening in the back wall of the fireplace that tapered from the outside in, and through which an air blast was introduced by means of a water-powered leather bellows. The tuyere, about seven inches above the floor of the fireplace, was constructed at a slight vertical angle in relation to the fireplace floor. It was this angle that determined the quantity of the iron produced. If the angle of the tuyere was nearly horizontal, a large amount of iron could be smelted from a small charge of ore, but at a high cost in fuel. As the angle increased from the horizontal, the quantity of necessary fuel decreased, but so did the amount of iron produced from a standard charge of ore.[24] It took a knowledgeable builder to determine the optimum angle for the tuyere.

With the bloomery constructed, the next step was to prepare the fuel, charcoal, from sound hardwood trees, preferably hickory or oak. Workers felled trees and split and cut the wood into four-foot lengths. Colliers then closely piled the wood in heaps with a small opening in the top for ventilation, covered them with twigs, leaves, and earth, and ignited the center of the heap. After almost a day had elapsed they closed the top and the wood smoldered for three or four days. The heap slowly settled, and when smoke stopped escaping the charred wood was left for four or five days to cool. Finally workers raked off the covering and carted the charcoal to the bloomery.[25]

In primitive bloomeries the floor of the fireplace was made of stone, and it was necessary to warm it gradually to prevent it from splitting. To make iron, iron workers first heated the bloomery fireplace with a small charcoal fire, and they usually lined it with charcoal dust before the fire was lit to make the warming process more gradual. Once the fireplace was thoroughly heated, they added pulverized iron ore to the top of the burning charcoal and slowly increased the velocity of the air blast coming through the tuyere. As the ore began to melt, the liquid iron ran down through the charcoal to the bottom of the fireplace, while workers added more charcoal and ore to the top of the burning pile. Workers shaped the molten iron, settling and cooling at the bottom of the fireplace, into a globular, semimolten mass called a bloom. When about one hundred pounds of iron thus had accumulated, in a bloom that was four or five inches in diameter, workers removed it from the fire with long crowbars and tongs. The product was ready to be hammered into bar iron.[26]

If the quality of the ore was good, and the blast and angle of the tuyere correct, about twelve hundred pounds of blooms could be made in a twelve-hour day. To make a ton of iron—a standard measure—took

Tilt hammer.

"The most simple machine by which iron is forged," Frederick Overman wrote, "is the German forge hammer, often called the tilt hammer." A fifty-pound hammer could forge nails, but Overman recommended a four-hundred-pound hammer *(a)* to convert sixty-to-one-hundred-pound pigs *(c)* into bar iron.

almost two days and consumed three tons of ore and three hundred bushels of charcoal.[27] There was enough work for everyone—the bloomery operator, the miners, and the colliers.

After lifting a bloom from the fireplace, workers carried it to an anvil where it was hammered into bar iron, either by several men with hammers, or by means of a water-powered tilt hammer. The head of a tilt hammer weighed between 250 and 600 pounds and was attached to a long wooden helve, or shaft, secured to a fulcrum. A great waterwheel turned a smaller wheel with metal wipers that struck the tail of the helve and tilted the hammerhead up off the anvil. Recoil and gravity brought the hammer down with tremendous force at a rate of between eighty and three hundred blows per minute, depending upon the hammer's weight and the number and spacing of the wipers.[28] Constant hammering drove out the impurities in the iron and increased the metal's strength, but since blooms could be hammered only when hot, workers made many trips from the anvil to the bloomery fire and back, carefully reheating and hammering the iron. This resulted in a great expenditure of time and manpower, and a corresponding increase in manufacturing cost.

Operating expenses were high, and on 22 July 1773 Donelson mortgaged his home plantation by a deed of trust to Hugh Innes. The mortgage included 1,019 acres on the Banister River and eighteen slaves, and served as security for a loan of £1,000 from Innes to Donelson.[29] The loan was to be repaid over two years in bar iron valued at £25 per ton, in eight equal payments of five tons each, due every three months. The twenty tons of iron a year probably accounted for almost half the annual production. Donelson evidently paid his debt on schedule, since there are no further court records concerning it, but this transaction shows the seriousness with which Donelson approached the iron business. Although the loss of his home and slaves would have been a disaster, Donelson was willing to risk them to establish his industry, but he did not mortgage the ironworks property or the slaves who worked there.

The same thing that brought success to the bloomery—the settlement of the Franklin County area—must have made John Donelson discontented. By the beginning of the revolutionary war the Blue Ridge no longer marked the edge of the frontier, which now lay to the south and west, through the Cumberland Gap to Kentucky and southward to the Tennessee River. Donelson, after all, was possessed of that same adventurous spirit that compelled Daniel Boone, Simon Kenton, and scores of other men to lead civilization into the wilderness.

His duties as surveyor had permitted him to whet his appetite for exploration. In 1770, when John Stuart, superintendent of Indian affairs for the southern department, was about to conduct treaty negotiations with the Cherokee at Lochaber, in South Carolina, the governor and Council of Virginia proposed to send a representative on behalf of the colony. On 6 August 1770 they decided that "Col. John Donelson, of Pittsylvania County, is a proper Person for that Appointment, and also to be the Surveyor to run and mark such Line as may be established by the said Treaty."[30] In 1771 the boundary line was run and established the Kentucky River as the western boundary of Virginia.

While leading the surveying party, Donelson had crossed a vast tract of land seldom seen by white men. His desire for new land led him to associate himself with the Loyal Company, which was composed of speculators largely from the Albemarle County area. The Loyal Company had received grants of land in Virginia's Trans-Appalachian frontier totaling 800,000 acres between 1749 and 1752.[31]

Donelson may also have wanted to travel on because his family was breaking up. His eldest daughter, Mary, and her husband, John Chaffery, had decided to move to Louisiana, and his second son, John Donelson, Jr., was preparing to settle in Kentucky.[32] John Donelson himself had been selected by Richard Henderson, who unsuccessfully had tried to establish a new colony called Transylvania in eastern Kentucky, to lead a party of settlers down the Holston, Tennessee, Ohio, and Cumberland rivers to French Lick, the future site of Nashville.[33]

Another important factor in Donelson's decision to leave may have been a sudden setback in his iron industry. Marquis James, in his biography of Donelson's son-in-law, Andrew Jackson, wrote that in 1778 Donelson's speculation in an ironworks in Henry County "swept away the accumulations of thirty prosperous years."[34] Perhaps Donelson's iron market collapsed when the revolutionary war temporarily interrupted the flow of westward migration. Or possibly Donelson found that he could not make a profit on the limited amount of bar iron his bloomery produced, given the high cost of production. An increase in production and a lowering of overhead required the construction of a blast furnace, an enterprise for which Donelson was not prepared financially or technically. At any rate, Donelson seems to have preferred the role of frontiersman to that of entrepreneur. By 1779 he was ready to move on.

Having decided to sell the ironworks, Donelson quickly found two buyers from Bedford County, Colonel James Callaway and his father-in-law Colonel Jeremiah Early, who possessed the money and experience to enlarge the operation and turn a profit. Indeed, Donelson may have leased the property to Callaway and Early before the purchase. On 17 June 1779 Donelson executed a deed in which he sold to Callaway and Early for £4,000 the upper tract on Iron Mine Branch, consisting of 400 acres, and another tract he had purchased from Francis Byrd in 1775, containing 200 acres.[35] Donelson also transferred to them his claim to the inclusive survey of 3,988 acres that he had entered in 1773.[36] The ironworks now began the second phase of its history.

3

From Bloomery to Blast Furnace

Like John Donelson, Jeremiah Early and James Callaway were newcomers to the southern Virginia piedmont. Early, born in 1730 in either Middlesex County or Spotsylvania County,[1] had moved first to Augusta County and then to Bedford County. He advanced steadily in county politics, becoming a justice of the peace, sheriff, and colonel in the militia.[2] Callaway, born in 1736 in Caroline County,[3] moved with his parents to Bedford County in the 1750s. Although it was his father who led the way, James Callaway followed closely in his footsteps, succeeding him as a member of the House of Burgesses for the sessions from 1766 through 1768.[4]

In 1770 James Callaway formed a business partnership with Alexander and Peterfield Trent, of Rocky Ridge, in Chesterfield County.[5] Callaway sent hogsheads of Bedford and Pittsylvania County tobacco to the Trents at Rocky Ridge, across the James River from Richmond. The Trent brothers then shipped the tobacco to the firm of Dodson, Daltera, and Company, in Liverpool, England. There the tobacco was sold and the money was used to buy dry goods, which were shipped back to the Trents' store at Rocky Ridge, and to the branch store established in Bedford County by Callaway. In their first letter to the Liverpool merchants, the partners wrote about the Bedford store, that "where we fix this store is a large extent of Country, the lands fresh and good for Tobacco, the planters clear of Debt, and no store to Interfere with us."[6] The Trents and Callaway continued their business until 1775, when the Revolution stopped American trade with England.

Although the war ended Callaway's career as an international merchant, it quickly provided other opportunities for his talents. By 9 February 1776 he had been appointed superintendent of Chiswell's Lead Mines in Fincastle (later Montgomery) County.[7] The lead mines had been developed in the early 1760s, and when the revolutionary war began, they were taken over and operated by the state to furnish lead for the army. During Callaway's tenure as superintendent he was responsible for overseeing the mining, smelting, casting, and shipping of lead to places as far away as Pittsburgh. After two years as superintendent, Callaway decided to return to Bedford County and private enterprise. On 5 December 1777 "the Governor and Council appointed Charles Lynch Esquire to Superintend the managing of the Lead Mines in the room of James Callaway who is about to resign."[8]

After he left the lead mines, Callaway may have become a partner with David Ross, of Petersburg, in the operation of the Oxford Iron Works in Bedford County. Ross had acquired the works by 1776. In a letter written in October of that year from the Oxford Iron Works to John Hook, a local merchant and loyalist in New London, Ross told him "you may send down on Tuesday or Wednesday for your Iron."[9] In another letter written on 30 November 1776, Ross informed Hook that "my Smith has been so engaged at the Saw Mill for a few days that he has not done your Matock. I intend him to do the whole for you on Monday."[10]

The association of Callaway and Ross is based in part on the local tradition that Callaway built and operated the Oxford Iron Works during the Revolution.[11] This is not completely true, since Ross owned the ironworks when the war began; there is a strong tradition, however, that Callaway constructed a blast furnace on Ross's property, and some documentary evidence supports this tradition. On 16 April 1778 John Conway received from Charles Lynch at the lead mines "three Pounds ten shillings for going to Col. Callaway's Iron Works for Iron, etc., for the Use of Lead Mines."[12] On 1 December 1778 the public foundry at Westham near Richmond paid for nine tons, seven hundredweight, and forty-eight pounds of pig iron furnished the foundry by "Callaway and Early."[13] Since these payments predated the purchase of Donelson's bloomery by Callaway and Early, it is possible that both men were associated with Ross at the Oxford works. Because Ross was involved with many enterprises, he may have contracted with the two Bedford County residents to operate the works for him when he was in Petersburg tending to his other businesses. Certainly Callaway was a logical choice for such an undertaking because of his experience as a merchant and as superintendent of the lead mines. Thus, it may well have been Callaway, familiar with the operation of the blast furnace at the lead mines, who constructed the Oxford blast furnace for Ross.

On the other hand, the lead mine and public foundry payments may involve John Donelson's works rather than Ross's, since it appears that Donelson leased his property to Callaway and Early, and that the two partners actually built the Washington furnace while they were still operating under the lease. Early's son Joseph conducted the bloomery's business for Donelson, who was busy preparing for his expedition to

Tennessee. On 22 April 1779 a stray horse from Pulaski's Legion was turned over to William Bartee, of Henry County, who made out a receipt noting that he was "Two miles from Joseph Earley's Ironworks."[14] The most conclusive evidence, however, is an advertisement that appeared in the *Gazette of the State of South-Carolina* on 25 August 1779[15]:

> Washington Furnace
> *Henry County, in Virginia, July 21, 1779*
> Notice is hereby Given,
> THAT the said FURNACE entered into Blast the 1st of the present month, and it is expected will continue until Christmas next: where may be had, by wholesale or retail, the following assortment of castings, to wit, POTS, KETTLES, CAMP-KETTLES, OVENS, SKELLETS, FLATIRONS, SPICE MORTARS, FIRE DOGS, SMITHS ANVILS, FORGE ANVILS, FORGE HAMMERS, WAGGON BOXES, STOVES, or any other kind of Castings that may be wanted.
> Early and Callaway

If the furnace began operation on 1 July, its construction must have begun long before the sale of the ironworks by Donelson on 17 June.

It was fitting that Callaway and Early named their new furnace after the commander in chief of the revolutionary army, since the revolutionary war soon came to the partners' corner of Virginia. On 28 December 1778 James Callaway took the oath of office and assumed the duties of the county lieutenant for Bedford County, the same day that Jeremiah Early was sworn in as colonel in the Bedford County militia.[16]

As county lieutenant, Callaway was commander in chief of the Bedford County militia—a position of considerable authority, particularly in wartime. In 1780, when loyalists in the southwestern counties of Virginia, including Bedford and Pittsylvania counties, conspired against what they considered to be an illegal rebel government, Callaway and other civil and military leaders in the area took steps to suppress them. One of Callaway's partners in the effort, Charles Lynch, saw his last name become a byword for their actions: Lynch's Law. Those who were affected by it were forcibly imprisoned without trial, not hanged, and after the war Lynch, Callaway, and the others were protected from all liabilities arising from their actions by a special act of the General Assembly.[17]

The end of the war enabled Callaway to turn his full attention to his business affairs. Jeremiah Early, Callaway's partner, had died in the summer of 1779, shortly after the purchase of Donelson's bloomery. In his will, Early left his share in the partnership to three of his sons—Joseph, John, and Jubal Early—and appointed Joseph Early the agent "to Transact all the Bussiness of the said Iron Works, who shall be Accountable to my other two sons for their proportion of the Annual

Profits of the said Estate."[18] The inventory of Early's estate included 21 horses, 137 head of cattle, 6 of which were at the ironworks and were probably oxen, 210 hogs, 43 sheep, 62 slaves, plantation tools, a still, furniture, kitchen utensils, and 21 books, including Stack's *Justice, Joseph Andrews, Don Quixote,* and a dictionary. The estate was appraised at more than £86,000, excluding land.[19] At the time of Early's death, the ironworks consisted of the furnace, forge, outbuildings, and what he and Callaway had contributed to the partnership: "Slaves, Stocks of Horses, Hoggs and Cattle, Waggons and Gears, Plantation Utensils, Black Smith and Carpenters Tools, and all kind of Apparatus necessary for Carrying on and Erecting a Sett of Iron Works."[20]

After Early's death the task of enlarging the land holdings of the ironworks and building a larger forge fell to James Callaway, with the assistance of Joseph Early, aged twenty-three, John Early, twenty-two, and Jubal Early, fifteen. Joseph Early, however, died in the fall of 1780. In his will, datelined "Washington Iron Works, July 22, 1780," Early left all of his estate, including his share in the ironworks, to his brothers.[21] The inventory of his estate listed 25 tons of iron valued at £25 per ton, 7 slaves, 12,500 pounds of tobacco, an unspecified number of cattle, clothing, and a gun.[22]

On 23 March 1781 John Early sold his interest in the ironworks left to him by his father (retaining the share left to him by Joseph Early), to James Callaway.[23] John and Jubal Early then owned a one-third interest in the business, while James Callaway's share was two-thirds.

The blast furnace constructed by Callaway and Early was a complex and expensive piece of technology compared to the bloomery. Any expense and trouble, however, were repaid by increased efficiency and production. The bloomery had been well suited for the role it had played in supplying iron to the frontier inhabitants of southside Virginia. But the bloomery's production was limited to 40 to 100 tons of bar iron annually, in contrast with as much as 600 tons from a furnace and enlarged forge. In addition, bloomery iron was of uneven quality, "a mixture of fibrous iron, cast iron, and steel—an aggregation of unavoidable irregularities."[24] Finally, bloomery iron could not compete in price with the large quantities of iron produced by a blast furnace, because the cost of labor was too great in relation to the amount of iron produced.

Several factors had to be considered before the furnace was built. The site of the furnace had to be dry, close to a reliable source of water, and near a low hillside. It was important to build near the mines to keep the cost of cartage as low as possible. Such a location was found on or near the site of the old Donelson bloomery. Granite was quarried from the outcropping on Scuffling Hill just across Furnace Creek, and a foundation about twenty-eight feet square was dug and filled with rock. Slowly, as the tedious process of quarrying went on, the furnace rose from its foundation to a height of thirty feet, tapering toward the top.

More important even than the exterior was the interior construction of the furnace. The central cavity, called the bosh, opened from a diameter of three feet at the top of the furnace downward to a maximum diameter of about ten feet, and was lined with firebrick. The bosh narrowed again as it approached the base of the furnace until it formed the crucible, about two feet in diameter, through which the molten metal would flow down to the hearthstone. Three openings were made in the crucible. On the east side was the tuyere for applying the blast, and on the south side, at the bottom of the crucible, was a taphole with a plug of clay that would be punched out with an iron pole to allow the molten iron to run out onto the casting floor. Near the taphole was a larger opening that was blocked with a dam stone when the furnace was in blast. Occasionally, workmen removed the dam stone so they could clean out the crucible.

In the south or front wall of the furnace was a recessed opening, called the work arch, which was twelve feet high, ten feet wide, and five feet deep. In front of the work arch, protected by a shed roof, lay the casting floor, a thick bed of sand with a network of channels dug into it. The long channels going away from the work arch were known as sows, and the shorter side channels were called pigs. Because of the buildup of heat under the roof when the iron was flowing, a narrow vent extended from the top of the work arch to the top of the furnace to allow some of the heat to escape.

About fifty yards up the creek from the furnace stood an earthen dam from which a sluiceway led to an overshot waterwheel built between the furnace and the hillside. Water pouring over the wheel powered the great wood and leather bellows that was inserted into the tuyere notch through an archway ten feet high, eight feet wide, and four feet deep in the eastern face of the furnace.

On top of the furnace around the chimney opening ran a low rock wall a foot high, and a wooden fence probably bordered the edge of the furnace top. A step about a foot in depth and almost three feet from top to bottom was set into the top rear of the furnace. Here rested one end of a wooden bridge that spanned the fifteen-foot gap between the furnace and the hillside behind it.

Once the furnace had been completed, certain procedures had to be followed before it could be put into operation. First of all the iron ore had to be roasted with charcoal to burn away some of the impurities so that the iron could be more easily melted in the furnace.[25] To prepare for the roasting process, workers cut and charred great quantities of wood and carted iron ore from the mines to the furnace area where they broke it into pieces three or four inches in diameter. After clearing a circular

The busy eighteenth-century ironworks depicted in this conjectural drawing is the Marlboro furnace operated by Isaac Zane, Jr., in Frederick County, from about 1768 to 1790. Tub bellows above the waterwheel forced blasts of air down through the tuyere arch to the bottom of the furnace. On the casting floor in front of the work arch, a worker prepares trenches that will shape the molten iron into sows and pigs. In the foreground to the left are molds for casting firebacks and stove plates. The wooden shed surrounding the furnace sheltered workers from the weather. This ink sketch by W. Stuart Archibald is preserved in the collection of the Museum of Early Southern Decorative Arts, Winston-Salem, North Carolina.

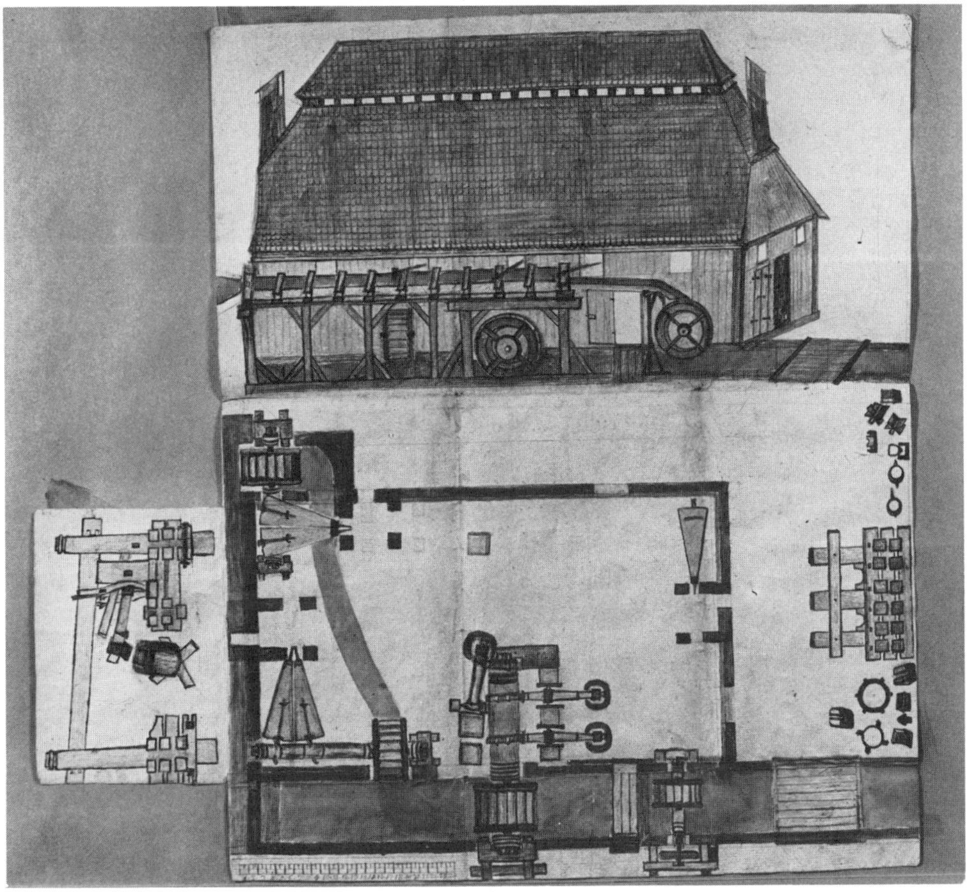

No colonial forge survives in the United States, but the Museum of Early Southern Decorative Arts preserves a 1797 drawing of a forge with four waterwheels, three fireplaces and bellows, and three hammers. The detail to the left shows a hammer frame and anvil. Photograph courtesy of the Museum of Early Southern Decorative Arts, Winston-Salem, North Carolina.

area on level ground, they laid billets of wood on the ground in a grid and filled the spaces between them with sawdust and wood chips. When the fuel reached a height of about eight inches, a two-foot layer of ore was piled on top, which in turn was covered with a two-inch blanket of charcoal. The workers continued to alternate fuel and ore until a height of six or eight feet was reached, and then they kindled a fire. The ore was roasted slowly, perhaps for as long as six or eight weeks, before it was ready for the furnace.[26]

The first firing of a new furnace took almost as long as roasting the ore. The ironmaster started a small fire on the hearth and kept it burning until the furnace stack was thoroughly dry. This was necessary because moisture could cause the rock to crack or explode if the furnace was subjected to intense heat before it was dry. After more than a month of drying the furnace was ready to be put into blast.[27]

Workers crossed over the bridge from the hill to the furnace mouth and poured in bushel basket or wheelbarrow loads of charcoal until the furnace was half full. Then they added the first charge of ore and limestone flux, the quantity of ore depending upon its quality. The ore charge was followed by a charcoal charge ranging from ten to fifteen bushels, then another ore charge, and so on until the fuel and ore almost reached the top of the furnace.[28] As the charcoal melted iron from the ore the contents of the furnace slowly settled. Workers were therefore employed around the clock continually charging the furnace with coal and ore.

As the furnace filled, a light blast was applied at the tuyere and slowly increased over the next two weeks until a full blast was reached. After perhaps a day in blast the first signs of iron, in the form of impurities called slag, appeared in the hearth. Soon after, molten iron ran down into the hearth and accumulated behind the dam stone. Workers removed the slag, which floated on top of the iron, and hauled it away after it had cooled. When the crucible was full of molten iron the ironmaster unplugged the taphole, and the iron ran down the sow channels into the pigs on the casting floor. Ironworkers scooped up some of the iron from the hearth in dippers and poured it into thick pottery molds to make a variety of cast iron and hollow ware, including stove plates, firebacks, kettles, skillets, and dutch ovens. Often workers cast large pieces such as stove plates and firebacks directly on the floor by pressing a carved wooden mold into the sand and then pouring molten iron into the depression. They broke pigs loose from the sows after they cooled and hauled them away to the forge to be hammered into bars.

The forge, which stood on the present site of the sewage treatment plant on Pigg River, about a mile from the furnace, contained two hammers, three refinery fireplaces for remelting pig iron, and one chafery fireplace for heating bar iron, all powered by waterwheels.[29] The founder carried pig iron to a refinery fire and heated it until it melted to a pasty mass, forcing out the impurities. He stirred the molten iron into a bloom and carried it with tongs to one of the hammers, where he laid it across the anvil. He manipulated the bloom so it could be hammered from the middle toward both ends, gradually forming it into a long bar with a large knob on the end where the tongs gripped it, and a smaller knob on the other end. At this stage the bar was known as an ancony. The founder reheated it in the refinery fire several times to sweat out impurities and keep it hot and malleable. To give the ancony its final shape as a bar, he carried it to the chafery fireplace and the smaller knob was heated. Once it had been hammered flat, the larger knob was given the same treatment.[30] The finished bar might be fourteen feet long, two inches wide, and half an inch thick[31] before it was cut into more convenient lengths for transportation and sale.

By the end of the revolutionary war, then, the Washington Iron Works had grown in size and complexity well beyond the limits of John Donelson's bloomery. In addition to the furnace and forge, a sawmill and gristmill were constructed.[32] To house the increased number of workers new dwellings were built, as were stables for the horses; sheds for storing ore, charcoal, and iron; a blacksmith's shop to shoe the horses and repair the tools; and an office for conducting the business of the ironworks.

All of this cost money, of course, though it is difficult to estimate how much. The initial purchase of Donelson's bloomery, mines, and 600 acres of land cost Callaway and Early £4,000. In addition, Donelson probably charged them for the transfer of his 3,988-acre inclusive survey, although this is not stated on the document. To these costs must be added those of constructing the furnace and forge, as well as the other buildings and equipment. In 1786 a Pennsylvanian estimated that "an iron plantation, including furnace, forges, gristmill and sawmill, and all equipment together with the necessary amount of land, would be cheap at £6,000 and dear at £12,000."[33] Given this estimate, it may well have cost Callaway and Early an additional £4,000 to build the furnace and forge, bringing the total investment to £8,000. Considering the high inflation of the late war years and the scarcity of skilled workers, however, this estimate may be too low.

In 1775 Virginia's legislature, recognizing the importance of iron manufacture to the war effort, had exempted from military service all "persons concerned in ironworks."[34] Another act, passed during the May 1777 session of the General Assembly, exempted "horses, waggons, or waggoners, employed at any lead, copper, or iron works,"[35] from impressment. At the May 1780 session the exemption from service granted to ironworkers was repealed,[36] but in 1781 the exemption was reinstated together with the immunity from impressment granted to wagons, wagoners, and horses,[37] to remain in effect through the end of the war.

The annual production of the Washington Iron Works during the late war years was equal to that of most other Virginia ironworks. Thomas Jefferson in his *Notes on the State of Virginia* wrote of the various works,

> Callaway's, Ross's, Millar's, and Zane's, make about 150 tons of bar iron each, in the year. Ross's makes also about 1600 tons of pig iron annually; Ballendine's 1000; Callaway's, Millar's, and Zane's, about 600 each.[38]

According to Jefferson's figures, Callaway and Early's forge produced about 25 percent of the bar iron, and their furnace about 14 percent of the pig iron, manufactured in Virginia. Bar iron and castings usually sold for £25 per ton; because of inflation, however, by 1781 the price had risen to £40 per ton,[39] and in 1782 it stood at £50 per ton.[40] The ironworks thus earned a gross annual income from the sale of finished iron products during the war years of £3,750 to £7,500. The cost of production must have increased proportionally, however, and no doubt the cost of skilled white labor went up even more. The end of the war and the slow return to normal prices must have been as great a relief to the entrepreneur as to the consumer.

4

James Callaway and the Washington Iron Works

The successful conclusion of the revolutionary war brought peace and stability to the southwestern piedmont of Virginia. The populations of Henry and Bedford counties increased as new settlers arrived, and many residents thought it was time for the formation of a new county. As early as 1779 those in favor of dividing the old counties had petitioned the General Assembly, saying that long distances and too many rivers separated them from their courthouses. Those opposed to the plan belittled the others' complaints and warned of the financial problems that would result from diluting the local tax base. Between 1779 and 1785 twenty-three petitions both pro and con were submitted to the General Assembly.[1] Finally, during the session begun in October 1785, the General Assembly passed an act creating a new county from parts of Henry and Bedford "from and after the first day of January next . . . called and known by the name of Franklin" and specified that the first meeting of the county court would take place "at the house of James Callaway, at his iron works in the said county."[2] The court accordingly met on 2 January 1786 and the new justices were sworn in.[3]

According to local tradition the first court meeting was held in what is now the home of J. Francis and Laquita Amos, overlooking the Washington furnace from a hill to the northwest. The earliest part of the present house is the front, which originally was a story and a half with two rooms and a center hall on each floor. The court is believed to have met in the northeast room to the right of the center hall on the first floor. While it is not known whether James Callaway was present at the first court meeting (he remained a resident of Bedford County after the division), it is clear that he attended others and was jubilant about the formation of the new county. On 2 February 1786 he wrote to Ralph Smith, of Pocket Planta-

According to local tradition, the first meeting of the Franklin County Court occurred on 2 January 1786 in the room at the right of the front door to the original part of the ironmaster's house, then owned by James Callaway, Sr.

tion in Pittsylvania County, asking his friend to send him a cask of brandy. "Unless I get this Brandy," he wrote, "I am to be without any thing to Drink at our Franklin Court Monday next [9 February]. . . . P.S. Let the Cask be full if Possible."[4]

Callaway's house at the ironworks was an ordinary as well as a dwelling. On 24 September 1784 the Henry County Court had granted him a license "to keep ordinary at his Tenement, near his and Early's Iron Works."[5] The justices of Franklin County, while meeting at Callaway's house on 9 March 1786, granted him a license "for Keeping an Ordinary at his place."[6] Shortly thereafter, a permanent courthouse was constructed in Rocky Mount where the present courthouse now stands, and the following year, when it was time for Callaway to renew his ordinary license, he decided a location in town would be more profitable than continuing at the ironworks. He built or purchased a building in Rocky Mount, and on 8 March 1787 he was licensed to keep an ordinary "at this Courthouse."[7]

Callaway soon decided to expand his operation to include a general store and to replace his ordinary with a tavern offering lodging as well as meals and refreshments. The store came first. On 3 April 1787 the court granted a license to Daniel Brown, Callaway's son-in-law, "and Company" to keep a retail store.[8] The words "and Company" indicate that Callaway and Brown were partners in the store. When the license was renewed on 8 April 1788, the court clearly stated that "Daniel Brown and Callaway is Permitted to keep a Retail Store. Also leave is Granted [Callaway] to Keep a Tavern in Rocky Mount."[9] According to an insurance policy issued on 5 November 1805, the tavern and store were housed together in a one-story wooden building eighty-four feet long and twenty feet wide, with three chimneys and a wooden portico seventy feet long and ten feet deep.[10] The same policy also listed two dwelling houses, an office, and a dairy.

The steady growth of Callaway's various enterprises in Franklin County, dependent upon cordial relations with his resident partners who managed the daily affairs of the businesses, was threatened by his partner John Early, who sued Callaway in the General Court in March 1786. Early claimed that on 10 May 1784 Callaway ordered the manager of the ironworks to withhold from sale Early's one-sixth share of the bar iron made there, which at that time amounted to eight tons. In addition, Callaway insisted that Early pay his one-sixth share of the manager's wages, as well as his share of the ironworks expenses. This situation persisted for eighteen months, declared Early, and Callaway

> artfully and maliciously forbids and prevents the sale of the said Plaintiffs proportion of Iron that he may thereby force and compell the said Plaintiff to sell his said share for much less than its real value and has greatly injured the Plaintiff to his Damage.[11]

Callaway replied that he was innocent of Early's charges. The case was continued several times until April 1790, when it was dismissed, Callaway and Early having come to terms. One can only speculate whether or not Early's charges were true, and whether Callaway was as ruthless a businessman as the charges indicated.

The growth of Callaway's businesses in Franklin County increased his influence in the political affairs of the county. Some residents resented this, perhaps viewing Callaway as an "absentee landlord" running things from his home plantation in Bedford County. One in particular who held that view was the old loyalist, John Hook, whose store was located in Franklin County instead of Bedford County after the division. Perhaps Hook and Callaway had tangled back in the days when Lynch's Law had kept the tories in their place in Bedford County, and Hook still harbored a grudge. Regardless of the cause, Hook composed a petition to the House of Delegates on 8 October 1787 in which he charged that Callaway had influenced the selection of the site of the Franklin County courthouse for his own profit.[12] Hook's accusations serve to illustrate the importance of the ironworks to the county.

The boundary line of the new county, according to Hook, had not been run in compliance with the act creating the county, but "four or five Miles farther to the South West than was Intended." This had resulted in the "Manifest Injury and Inconvenience of a great Proportion of the Citizens of Said County . . . for the Immolument of an Individual." The courthouse not only adjoined the Washington Iron Works property, but it was

> very inconvenient to Water entirely without Pasture and very Scarce of Timber. There is only one Spring and that a considerable distance from the Seat laid off for the Court house and down a very long Steep Hill and no Wells. The Scarcity of Timber Joining Iron Works Lands is too obvious to be pointed at.

Hook went on to lament that the nearby countryside could not support the needs of the ironworks for grain and provisions, much less the demands of a courthouse village, and that the lack of foodstuffs had caused the shutdown of the furnace in 1787. Then Hook presented the heart of his complaint:

> The fact is that Immediately before the Court house was fix'd on to be on the Lands of Col. James Callaway Joining the Iron Works Land the Members Comprising the said Court before they convened in the business of fixing the Court house appeared to be unanimous in fixing on the Centre of the County by an Actual Survey . . . but to our great Surprise and disappointment on Col. Callaway Petitioning to them for the Court house on his Rockymount Land where they were

> then setting (no doubt with great Eloquence from the Effect) their Worships in Compliance Acquiessed in fixing on the same for a Centre.

Hook urged the House of Delegates to pass an act requiring that a new boundary survey be made to determine the true center of the county and to order that "the Court house shall be removed and fix'd as near the Centre of the County as Convenience Admit."

Oddly enough, for John Hook was no shy or reticent man, he appears never to have submitted his petition, or else he withdrew it before action could be taken. Perhaps he merely was venting his spleen against Callaway, because the first courthouse had been built by the time he wrote his petition. He probably realized that no one would want to move the court now that it had settled into the courthouse, and, as a glance at a map would show, Rocky Mount was not so far from the geographical center of the county as to be a great inconvenience to anyone. Besides, as John Hook himself had predicted in his petition,

> the business of the Gentleman in question [Callaway] will in all Probability sooner or later be very Considerable in the said Court from his being largely in Trade in the said County and very rich and Powerful.

Being rich and powerful did not automatically protect James Callaway from every challenge. In 1790, after an absence of twenty-two years, John Wilcox returned to Virginia from North Carolina and laid claim to the tract of land containing the Washington Iron Works.

On 12 July 1790 James Callaway and the heirs of Jeremiah Early were issued a grant[13] for the inclusive survey of 3,988 acres made for John Donelson in 1773 and transferred to Callaway and Early in 1779. Apparently the partners did not know that John Wilcox actually had won his case against John Donelson in the General Court. Wilcox had filed a caveat in 1772 objecting to the issuance of patents to Donelson, and had been upheld by the court on 9 April 1783.[14] On 30 October 1784 grants had been issued to him for the two tracts of 400 and 403 acres on Iron Mine Branch.[15] These tracts were contained in Callaway and Early's inclusive grant of 3,988 acres issued on 12 July 1790. Obviously either Wilcox's, Donelson's, or the partners' grant was invalid.

Wilcox sued Callaway in the High Court of Chancery, seeking to obtain full title to the disputed land on the ground that he had been defrauded by John Donelson. Callaway claimed that he had purchased the land from Donelson in good faith and did not first hear of a possible cloud on the title until three years later, in 1782.[16] The court agreed with Callaway and ruled in his favor. Wilcox appealed the decision to the

Supreme Court of Appeals of Virginia, where the case was heard during the fall term of 1791. The court upheld the earlier decision, ruling that no fraud had been proven, but that although Wilcox could not receive full title he could claim with justice that he had a lien upon the land. Given the decision Wilcox and Callaway had no reasonable alternative but to settle the matter out of court. This they did on 24 November 1791, when Wilcox deeded to Callaway "all his right in law and equity" to the disputed land for £250.[17] At last James Callaway had a clear title to the Washington Iron Works.

Ever since his purchase of Donelson's bloomery in 1779, Callaway together with the Earlys had been adding to the ironworks property, both by obtaining land grants from the state and by purchasing land from individuals. In 1786 the firm of Callaway and Early owned 12,000 acres of land in Franklin County;[18] by 1791, when the suit with Wilcox was settled, the number of acres had grown to 13,171.[19] Callaway's personal land holdings in Franklin County, exclusive of the partnership, were 3,983 acres in 1786[20] and 13,342 acres in 1791.[21] In 1809, the year of Callaway's death, he owned 21,571 acres in his own right; the ironworks property consisted of 18,908 acres; and the estates of John and Jubal Early owned 6,171 acres, for a total of 46,650 acres controlled by Callaway and the Earlys.[22]

Most of the iron made at the furnace and forge, judging from the number of lawsuits involving Callaway and Early, was sold locally. Buyers came to the ironworks office for their iron, or bought it at country stores whose owners had purchased castings and bar iron from the company. Callaway, however, was determined to expand his market. In a letter written from the ironworks on 25 October 1787 Callaway discussed where he proposed to sell his iron, besides in the local market: "I shall send a part of it to the Moravians, the Balance to the most Probable Places for a Market, Perhaps a part, Sixty or Eighty Miles down the Country."[23] In his reference to the Moravians, Callaway probably meant the large settlements of the German sect in North Carolina, such as that at Old Salem. Perhaps the ironworks market had contracted because of competition after the revolutionary war, during which his iron was sold as far away as South Carolina.

Transporting iron to distant markets in five-horse wagons was a slow and costly means of travel. Since the ironworks foundry was located on Pigg River, water transportation would have been cheaper and faster had the river been navigable. Hoping to obtain the blessing of the state, Callaway and Early addressed a petition to the legislature in 1796 to authorize the improvement of Pigg River:

> The Petition of Callaway and Earleys, Humbly shew that the improving the navigation of Pig River from the mouth of the said River to

> Washington Iron Works in the County of Franklin, would be attended with great convenience to your pe[ti]tioners and a number of other Citizens who live convenient thereto. Your Petitioners have made an experiment on the said River, by a Boat with Eighteen or nineteen hundred weight of Iron which was taken in and transported from the said Forge to Staunton River with but little obstruction.[24]

The company then asked the General Assembly to authorize the improvement of the river by private subscription and to appoint commissioners to oversee the work. On 13 December 1796 the General Assembly agreed to the petition and passed an act appointing as trustees Swinfield Hill, Benjamin Cook, Josiah Woods, Samuel Duval, and James Callaway, Jr.[25] The trustees were empowered to clear and improve the river so that boats carrying three thousand pounds of goods might pass unimpeded.

Not every inhabitant of Franklin County was enthusiastic about the prospect of river improvement, however. Lewis Burwell, whose mill on Pigg River had been destroyed by accident, claimed in a 1797 petition to the General Assembly that the river improvement act had prevented him from rebuilding the mill. He doubted, he wrote, whether Pigg River or Staunton River could ever be cleared without enormous expense, and asked that the act of 1796 be repealed. Furthermore, Burwell asserted,

> if the carrying of iron down the river was one of the objects contemplated, your petitioner is warranted in saying that the demand by land has been more than equal to the supply. And altho' it is painful even to suggest improper motives, yet your petitioner must be pardoned for hazarding a conjecture, that the probability of a rivalship from a Furnace and Forge at his mill seat, where his land abounds with iron ore, may have been a greater stimulus to some, than the public good. And whether that interest, in the present state of things, will not be more effectually promoted by a Furnace and Forge where nature seems to have intended works of the kind, he willingly submits to the determination of the General Assembly.[26]

A second petition filed on the same day as Burwell's, and written by the same hand, likewise stated that

> if the transportation of iron is offered as an inducement [to river improvement], it may with truth be observed that the supply of that article at a high price has hitherto been short of the demand by land and probably will continue so until other works shall rival those already erected.[27]

This petition was signed by twenty-four citizens of Franklin County who lived near Burwell's mill; it, like Lewis Burwell's petition, was rejected by the General Assembly.

Despite the current of opposition, most inhabitants continued to support the idea of river improvement and the dream of increased trade opportunities. In 1799 an enthusiastic group of Franklin County citizens petitioned the General Assembly for a law, similar to that passed in 1796, to improve Staunton River and Blackwater River. The petitioners sought to have Blackwater River rendered navigable from its mouth on Staunton River up to the forge owned by Swinfield Hill, Walter Bernard, and William Armstrong. Hill and Company owned a blast furnace on Story Creek about seven miles southwest of the Washington Iron Works, as well as the Blackwater forge; the forge and furnace, known as the Carron Iron Works, provided the only competition in the county for Callaway and Early. Perhaps Hill and Company furnished the inspiration for the petition; Swinfield Hill's signature led all the others. The petitioners pointed out that if the rivers were improved,

> they would be enabled to Carry to Market the produce of their farms at a much less expence and more expedition than by Land through a hilly Country as is the present mode at an enormous expence. . . . Agriculture and Commerce would flourish amongst us.[28]

The petitioners' dreams, however, foundered upon hard reality. The cost of dredging and deepening a river, which had to be done by manual labor, was huge. Apparently neither the petition of 1796 nor that of 1799 raised enough money to begin work. On 3 January 1804 the General Assembly passed an act repealing the Pigg River navigation act of 1796.[29] The repeal was in response to a petition submitted on 16 December 1803, which stated that "many Years has elapsed since the passage of the Law aforesaid without any attempts being made either to improve the Bed of Pig river or navigate the same."[30] The reasons offered for the repeal of the law were plainly economic: Robert Harvey, of Botetourt County, had dammed Pigg River about eight miles downstream from the Washington Iron Works and was constructing an ironworks of his own there. The petitioners claimed that

> the demand for Bar Iron at James Callaways Works had been for many years greater than the Works could Supply from Several Counties in Virginia and North Carolina, said Callaway Having a Monopoly of that Business in the County of Franklin except a Small Supply that comes from Robert Harveys Furnace in Botetourt. Your petitioners Conceive that a Publick Good will Result from the Iron Business being owned and Carried on by different owners. The Supply will be more certain and payments made with more ease from the produce of our Soil.

James Callaway's monopoly of the Franklin County iron industry was, as the petitioners pointed out, virtually complete by the time the Pigg

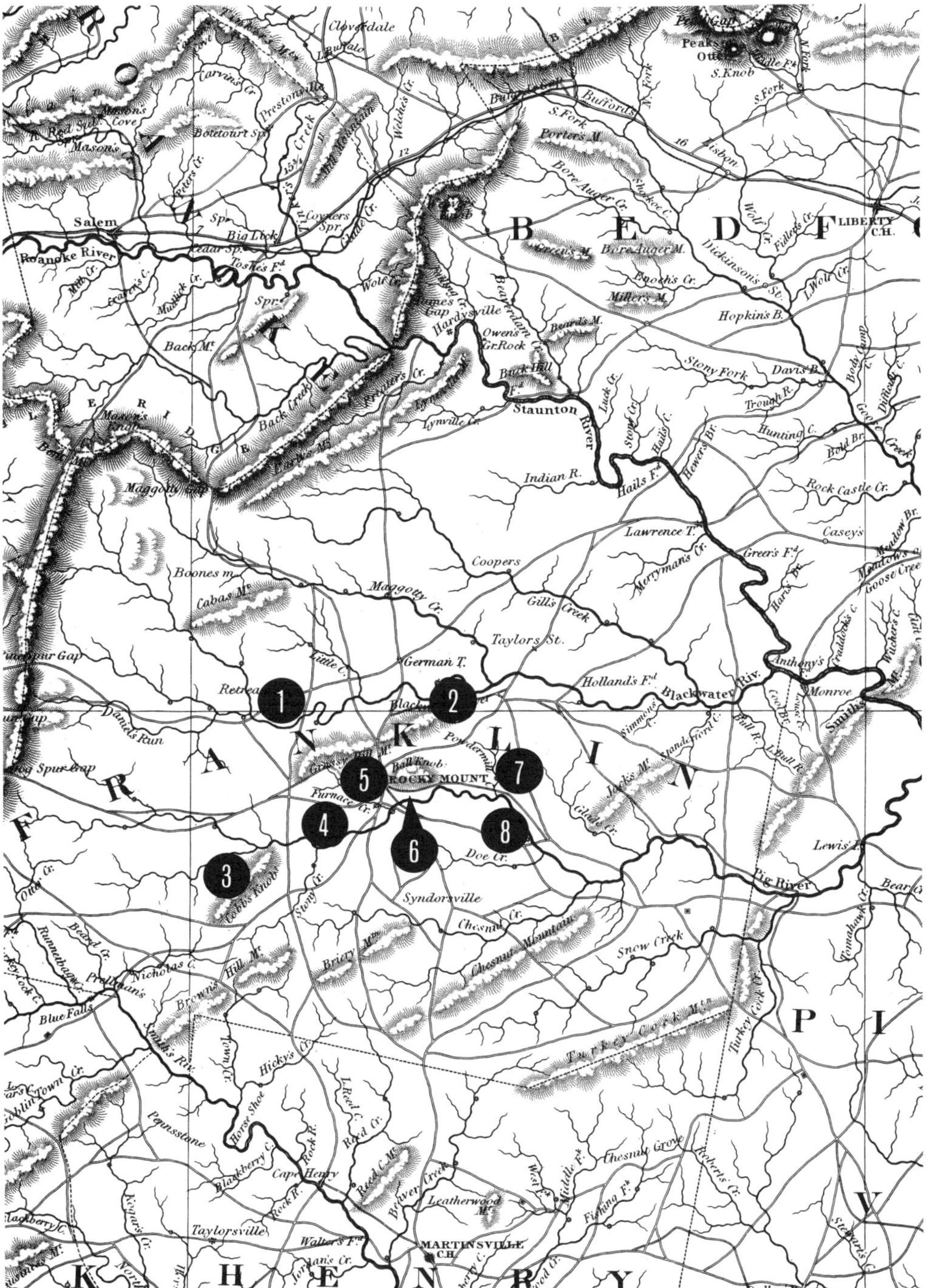

Iron-manufacturing sites in Franklin County between 1781 and 1865 are located on this detail from the Böÿe-Buchholtz *Map of the State of Virginia* (1859): (1) Elk Forge, (2) Hill and Company forge, (3) Carron Furnace, (4) Valley Forge, (5) Washington Iron Works furnace, (6) Washington Iron Works forge, (7) tilt hammer tract, and (8) Robert Harvey's forge.

River navigation act was repealed. Swinfield Hill, Walter Bernard, and William Armstrong, owners of the Carron Iron Works furnace and forge, had become indebted to John Hook in 1801, and had mortgaged the ironworks. David Ross, formerly a partner of Hook and in 1801 involved in an acrimonious lawsuit against him, had obtained an injunction stopping Hill and Company's payments to Hook. This injunction prevented Hill and Company from fulfilling their obligation to Hook, and their ironworks property was ordered sold.[31] The purchaser was James Callaway. In 1801 he bought Hill and Company's ore banks at the head of Town Creek,[32] and on 2 April 1802 he acquired the furnace and sixty acres of land on Story Creek.[33]

As owner of the Washington and Carron furnaces, Callaway enjoyed total control of the production of pig iron in Franklin County. By 1804 he also owned two of the four forges operating in the county (the Washington and Carron forges); his only competition came from Robert Harvey's forge under construction on Pigg River, and from the Elk Forge owned by John Miller, of Botetourt County. This forge was in operation by 1795,[34] and was located on Blackwater River upstream from the Carron forge. An 1800 newspaper advertisement pointed out that besides the forge, the property contained a gristmill and a sawmill and was considered very valuable "on account of the convenience to the Carolina trade."[35] The forge continued to operate independently of Callaway until it was closed between 1810 and 1820. Callaway's domination of the Franklin County iron industry was so complete that neither the Elk Forge nor Harvey's forge seriously threatened it.

The successful operation of an ironworks required more than buildings, land, and a ready market. It required labor, and in antebellum Virginia, that meant slave labor. Because of his wealth and his acute business sense, James Callaway had available to him from the earliest days of the ironworks a steady pool of slave labor from which to draw the most capable workers. When he and Jeremiah Early formed their partnership each was able to contribute to the ironworks labor pool because of their wealth. They might not have been able to expand as quickly as they did the scope of the ironworks operation (a labor-intensive undertaking), however, except that they seized an unusual opportunity that presented itself early in 1779.

In March of that year Robert Walton, Jr., of Georgia, fled from his native state in the face of the advancing British army to some of his family's landholdings in Virginia. He brought slaves with him, young and old, some of whom he had inherited from his father. When Walton arrived in Virginia almost destitute, he sought a loan and Callaway and Early agreed to lend him what he needed. On 27 April 1779 they signed a contract providing that in return for a security deposit of thirty-one slaves Walton received the loan of £12,304 for eight months, to be repaid with interest. In addition, Callaway and Early allowed Walton a credit to be applied to the outstanding balance of the loan of £8.6.8 per month on each slave deposited. The agreement provided that in lieu of cash, Callaway and Early could choose to accept payment of the debt "in Slaves at 900£ each for Likely young fellows and Women."[36]

When Walton proved unable to repay the loan he was prepared to assign ownership of selected slaves to Callaway in accordance with the contract. Both parties, however, found that they were involved in an illegal transaction because, unknown to them, the General Assembly had passed an act during its October 1778 session forbidding the importation and sale of slaves. Walton, who had contracted further debts in Virginia, wanted to fulfill the obligations of his contract with Callaway and Early and return home to Georgia. Walton left Virginia in January 1783 but his slaves remained in the possession of Callaway. Through his lawyers in Virginia, Walton petitioned the General Assembly on 9 June 1783 for a special act enabling him to discharge his contract with Callaway and to sell as many slaves as necessary to satisfy his other debts. Callaway supported the petition, which was considered reasonable by the House Committee on Propositions and Grievances.[37] The committee, however, never reported its conclusion to the House of Delegates, and no act granting the petition was ever passed. Robert Walton, Jr., continued to pay the taxes on the slaves until 1786, when they disappeared from the tax rolls, perhaps because Walton brought them back to Georgia.[38] James Callaway, meanwhile, had worked Walton's slaves for almost three years.

The number of taxable slaves over twelve years of age that Callaway and Early's firm owned, rather than rented, grew from year to year. The company had 9 such slaves at the ironworks in 1783,[39] 11 in 1784,[40] 20 in 1785,[41] and 27 in 1786.[42] By 1801, when Callaway began acquiring the Carron Iron Works, the firm owned 38 taxable slaves at the Washington Iron Works,[43] and at Callaway's death in 1809 the company's slave force numbered 48.[44] In addition to these laborers, Callaway employed several free, white men in various occupations. Thomas Crutcher operated the tavern in Rocky Mount[45] after the death of Daniel Brown on 25 March 1795.[46] Crutcher had been orphaned in 1786 and had chosen as his guardian William Menefee,[47] who was a "founder at the furnace of Callaway and Early."[48] William Thompson worked as a "Forge Man" in the 1790s,[49] and B. Smith was the clerk in 1786.[50] James Traner worked as a clerk in 1793,[51] replacing Pratt Hughes, who was first the clerk, then the "mannager of the Iron Works for Callaway and Early's"[52] in the 1780s and 1790s. William Crump managed Callaway's store during the same period.[53]

James Callaway died in 1809, after having owned the Washington Iron Works for thirty years. In his will dated 2 May 1809 he left to his son James Callaway, Jr., "a small piece of Land called Rocky Mount with the

Nineteen and one-half inches wide at the handles, this iron kettle was manufactured at the Carron works, in Franklin County, which operated from about 1796 to 1820. Photographs courtesy of the Museum of Early Southern Decorative Arts, Winston-Salem, North Carolina, gift of Mr. and Mrs. Joe Milton.

Tavern and all other improvements thereon except what has been reserved for the County on which the Courthouse stands."[54] To Henry T. Callaway, John Callaway, and George Callaway, his sons; William Callaway, his brother; James Steptoe, his son-in-law; and Christopher Clark, his friend, all of whom were named executors of his will,[55] James Callaway left his shares in the Washington and Carron ironworks.

The estate of James Callaway was appraised at almost eleven thousand pounds,[56] exclusive of his landholdings. At his home in Bedford County, the Forest, were 22 slaves, 19 horses, 42 sheep, 19 cattle, 6 oxen, an unspecified number of hogs, 3 stills, and an old coach.[57] His household furniture was elegant, including a secretary desk and tea table, both inlaid, a sofa and a dozen mahogany chairs in the parlor, and a sideboard, two "large black walnut Tables," and 12 chairs in the dining room.[58] He also died possessed of "one Silver Watch."[59]

The inventory of the property in the Rocky Mount tavern gives an idea of its size. It contained 13 bedsteads, 42 split-bottomed and Windsor chairs (some damaged), and 7 tables. There were also 9 pairs of fire dogs, 4 large ovens, a small oven, and assorted kettles, skillets, tea kettles, and pot racks.[60]

At the Washington Iron Works were 7 slaves of all ages owned entirely by Callaway and 58 slaves owned jointly by Callaway and the heirs of John and Jubal Early. Equipment and tools used at the ironworks included 6 wagons, "new wood work for 2 Waggons," an ox cart, 12 plows, 15 hoes, a set of blacksmith's tools and bellows, 15 axes, a hammer, and several saws, chisels, and wedges. Livestock totaled 28 horses, 51 head of cattle, and 146 hogs. The house contained 3 beds, and there were the usual pots and plates in the kitchen.[61]

The Carron furnace and forge were appraised separately. At the furnace were 11 slaves, 11 horses, 8 head of cattle, 29 hogs, 2 wagons, and a "sorry Bed."[62] Thirteen slaves were listed at the forge, with 18 horses, 11 head of cattle, 67 hogs, an old wagon, and "new wood Work of a Waggon." The forge also had a dwelling house and kitchen containing 3 beds, a bookcase, iron dishes, and the usual kitchen equipment. Tools included axes, hoes, wedges, scythes and cradles, and a "broken sett of Smith's Tools."[63]

Within a year of Callaway's death the Carron furnace had ceased operations. In 1810 Tench Coxe, in his survey of manufactures in the United States, reported only one blast furnace in Franklin County, the Washington furnace, with an annual production of 300 tons of pig iron and castings.[64] Coxe also reported four forges in operation in Franklin County: the Washington forge, the Carron forge, Robert Harvey's forge on Pigg River, and the Elk Forge on Blackwater River. The total production of these forges was 470 tons of bar iron valued at $115 per ton.[65] The difference between the 300 tons of pig iron and castings produced by the

Washington furnace and the 470 tons of bar iron produced by the four forges probably was made up by using old inventories of pig and scrap iron, as well as by purchases from furnaces in nearby counties.

Managed by William Callaway, George Callaway, John Callaway, James Steptoe, and Christopher Clark, the Carron forge and the Washington Iron Works continued in operation after the death of James Callaway.[66] The managers were soon joined by William Langhorne, who had married Catherine Callaway, James Callaway's daughter, on 26 February 1811.[67] In 1812 Callaway's executors added to their company a thirty-six-year-old merchant and Rocky Mount postmaster named Peter Saunders.[68]

When John Donelson sold his bloomery to Callaway and Early in 1779, he did so in part because he wanted to follow the frontier as it moved westward to Tennessee. The industrial revolution had not yet arrived in America, but the demand for iron in 1779 required a different type of man to operate the furnace and forge, someone more settled in his ways than a frontiersman but nonetheless possessed of imagination. James Callaway, a gentleman planter-ironmaster, bridged the gap between the passing of the frontier in Virginia and the arrival of industrialism. In his life and character he had more in common with William Byrd II and Governor Alexander Spotswood than he did with the industrialists of the nineteenth century, but he had the drive and business sense of the entrepreneurs who came after him. Callaway's death exemplified the passing of the gentlemanly business world of the eighteenth century. If the Washington Iron Works were to continue to prosper in the new industrial age, it would need a man of that age, less of a gentleman and more of a businessman, to give it renewed vigor. Peter Saunders was that man.

5

Peter Saunders, Ironmaster

John Donelson, for all his activity, is a shadowy figure of the frontier, always on the move, owing little allegiance to a particular colony or state. James Callaway was well known in the southwestern piedmont region of Virginia and the adjacent counties in North Carolina, a settled and respectable man of wealth and influence who generally remained aloof from controversy. Peter Saunders, however, was almost entirely a man of Franklin County, having only an occasional personal influence beyond its boundaries, and so was involved intimately with the people and issues of his home county. In contrast with Donelson and Callaway, Peter Saunders was reckless and domineering—a lightning rod for controversy. His arrogance and single-minded determination to have his own way kept the political and social leadership of Franklin County in an uproar for most of his life.

His father, Peter Saunders, Sr., was born in Goochland County on 20 September 1748.[1] He married Mary Sparrel on 31 October 1767 and shortly thereafter moved to western Pittsylvania County where he settled on Runnet Bag Creek. Like Donelson and Callaway, the elder Saunders assumed positions of authority, becoming a justice on the court of Henry County when it was formed in 1777 and a lieutenant colonel in the militia in 1782. At the start of the revolutionary war he served on the committee of safety for Pittsylvania County, and he was a member of the House of Delegates during the war years.[2] When Franklin County was organized in 1786 he was one of the justices sworn in at James Callaway's house at the ironworks.[3] On 2 May 1786, Peter Saunders, Sr., was recommended to the governor as a "Proper Person to execute the Office as a County Lieutenant."[4] He became a merchant, obtaining a license to operate a retail store on 3 April 1787.[5]

Saunders and his wife had nine children, four boys and five girls.[6] Peter Saunders, Jr., was the eldest son, born 21 July 1776.[7] Two of Peter's

Founders cast firebacks *(top left)* and sows and pigs *(top right)* in sand molds at the furnace, and hammered pig iron into bar iron *(bottom left)*. They cast smaller iron objects and hollowware in earthenware molds and often tossed the waste pieces, called sprues *(bottom right)*, into nearby Furnace Creek.

brothers—Fleming, born in 1778, and Samuel, born in 1783[8]—lived to adulthood, shared with him in the ownership of the Washington Iron Works, married, and had children. Peter Saunders, Jr., never married.

A first-born son can receive special pressures to succeed, especially if he is the namesake of a successful father. In addition to his father's example, young Peter Saunders witnessed the success of his younger brothers. Fleming Saunders chose the law for his life's work, was educated at the University of North Carolina, and rose in his profession to become a judge of the Circuit Superior Court of Law and Chancery in Virginia. Samuel Saunders overcame the handicap of an amputated leg to become one of the leading planters and landowners in Franklin County, as well as a judge on the county court. Like their father, Fleming and Samuel were never touched by controversy, in contrast to Peter.

In 1799 or 1800 Peter Saunders, Jr., fathered an illegitimate daughter. There is no record of a grand jury presentment for bastardy and the identity of the baby's mother has never been determined. The fact of the child's illegitimacy is based upon two sources: Saunders family tradition and Peter Saunders's own admission. According to the family tradition, the mother appeared at Peter Saunders's door not long after the girl was born and thrust her into his arms, saying, "Here—she's yours, so you raise her."[9] Peter Saunders's own admission was made in 1840 when he wrote his will. In it he referred to "my natural daughter Jane J[ones] Saunders who now resides with me."[10] In the eighteenth century the word *natural* when applied to a child simply meant "my own child," and carried no imputation of illegitimacy. By 1840, however, the meaning of the word in that context had changed completely. In his dictionary of 1841 Noah Webster defined *natural* as "illegitimate; born out of wedlock; as, a *natural* son."[11]

Jane Jones Saunders lived in her father's household for most of her life, acting as his hostess and managing the domestic life of the plantation. She never married. The Saunders family accepted her as a member of the family and referred to her as "Cousin Jane."[12] She died at the home of John S. Hale in Franklin County "on the 5th day of December 1861 in the 63d year of her age,"[13] and was buried next to her father.

Despite his transgression, young Peter Saunders appeared likely to follow in his father's footsteps as a leader in Franklin County. In 1796 he was recommended to the governor as first lieutenant in a company of artillery to be formed for the county militia; James Callaway, Jr., was recommended as captain of the company.[14] On 14 April 1803 Peter Saunders was appointed a justice on the county court,[15] and he was sworn into office on 3 May.[16] Far from being content with these honors, however, Saunders soon angered a number of people by seeking appointment as colonel of the militia, a position for which he was qualified neither by age nor experience. His artillery company had been disbanded soon after it was organized for lack of a cannon with which to train, and Saunders and Callaway had been declared extra or supernumerary officers.[17] As such, Saunders kept his commission but had no active role to play in the county militia.

Just how Peter Saunders convinced the county court, then, to recommend him to the governor in 1808 for promotion to colonel is a mystery. There was nothing mysterious, however, about the cries of outrage from his fellow officers who were senior to him in both rank and experience. One of them summed up the feelings of all very well when he wrote, perhaps making an oblique swipe at Saunders's sexual indiscretion of 1799,

> that however reputable Mr. Saunders may be in private life, . . . his appointment will be a direct slur on all the officers of the militia in said Regiment, for surely of them many are to be found whose bravery will not be questioned and whose talents will put them on a footing with that Gentleman.[18]

The court's recommendation finally was withdrawn, another officer was promoted, and Peter Saunders quietly turned his attention away from the militia and toward the business of getting a living for himself and his daughter. For the time being he was content to further his career as a merchant and to act as a justice on the county court. From 1810 to 1815 he also served as the postmaster of Rocky Mount,[19] an office which later would involve him in yet another controversy.

In 1814 Peter Saunders's brother Fleming married Alice Watts, of Campbell County.[20] She recorded her impressions of Peter Saunders in a letter written to her sister on 26 September 1814:

> Peter Saunders spent yesterday with us, and only left here about an hour ago. He is one among the handsomest men I ever saw, but, between friends, very dissipated. It is a pity he had not married some ten or twelve years ago, for it might have been the means of correcting his great fondness for gaming.[21]

It would have taken a very strong-willed women to keep a man of Peter Saunders's temperament from gambling. Perhaps Alice Saunders was referring to the mother of Jane Jones Saunders.

In 1814, after he had worked for Callaway's heirs for three years, Peter Saunders became a partner with William Langhorne in the operation of the Washington and Carron ironworks. The year 1815 also saw a tax imposed by the General Assembly on the production and sale of iron, and the tax records provide the most detailed account extant of the ironworks' production. From 10 February 1815 to 10 February 1816, 513 tons of pig iron were made by the Washington furnace. During the same period about 29.3 tons of hollow ware and castings and 219 tons of bar iron were produced by the ironworks. Only about 3.1 tons, or roughly 11 percent, of the hollow ware and castings, and 76 tons, or 34 percent, of the bar iron were sold during the year.[22] Sales dropped to zero after 1 August 1815, which may be explained by the falling price of iron.[23] Perhaps the partners held back their iron from the market in hopes of a rising price.

The late fall and early winter were the periods of greatest activity in the production of iron, followed by late spring and early summer. Production probably ceased altogether during the extreme temperatures of summer and winter. Bar iron was made throughout most of the year, since the forge was indoors, although it too probably closed down during the hottest part of the summer.

The landholdings of the Washington Iron Works, as presented in the 1815 land tax books, had declined slightly from 18,908 acres to 17,096 acres,[24] a decrease of just 5 percent since James Callaway's death in 1809. In 1815 the ironworks was taxed for 58 slaves over sixteen years of age,[25] an increase over the 48 taxable slaves in 1809. The Washington Iron Works also paid taxes on 22 horses, 40 cattle, and a book press.[26]

During the years from 1812, when he first worked at the ironworks, to 1815 when he became a partner in its operation, Peter Saunders had learned the intricate art of managing an iron plantation under the tutelage of William Langhorne. By 1817 he no longer was content with merely sharing in the management—he wanted to be an owner. To purchase the ironworks outright was beyond his means, so he called on his brothers, Fleming and Samuel Saunders, to assist him.

On 21 May 1817, William Langhorne sold to Fleming and Samuel Saunders "all the interest which he the said Langhorne has in the Washington and Carron Iron Works and in all the estate appurtaining to the same, real and personal, as the same stands and is held at this day."[27] The price was not low: $2,000 to be paid to Langhorne in ninety days, five tons of bar iron, $20,000 payable in installments over the next four years, and payment of Langhorne's share of a debt owed to the Bank of Virginia by himself and Peter Saunders, the amount not stated. In return for relieving Langhorne from all debts that he might owe to the ironworks, Fleming and Samuel Saunders received all his interest in money owed the ironworks by its customers. In the articles of agreement Langhorne described himself as the manager of the Washington Iron Works. His sale to the two Saunders brothers now put the management of the ironworks under the control of the Saunders family, although ownership of the property remained in the hands of James Callaway's executors and the Early family.

Even before this purchase, Fleming Saunders had been involved briefly in the manufacture of iron products. By 1814 a tilt hammer for fashioning tools from bar iron had been erected by Joseph H. Townes and Company on the property of John and Mary Callaway on the north side of Pigg River near Powder Mill Creek. The company placed advertisements in the Richmond and Lynchburg newspapers in the summer and fall of 1814 to

> respectfully inform the public, that they have just completed this establishment in Franklin County, Va., about four miles below the court house, immediately in the neighborhood of Callaway's Iron Works. They have provided themselves with good and faithful workmen, and the best materials for making Scythe-Blades, Reaping-Hooks, Weeding-Hoes, Drawing and Cutting Knives, &c., &c.
>
> They will also carry on the Waggon making business extensively—the wood and iron part of which will be made by neat and experienced hands. Orders and requests from any part of the Country, shall be punctually attended to.[28]

A price list followed, quoting, by the dozen, scythe blades at $24 to $30, reaping hooks at $7 to $8, weeding hoes at $8 to $9, drawing knives at $6 to $7, and cutting knives at $16 to $18. "Waggons, ironed complete," were $130 each.

On 20 December 1814 John and Mary Callaway sold the land on which the tilt hammer sat to Joseph H. Townes and his partner Alexander M. Hepbourn.[29] The property was mortgaged on the same day by Townes and Hepbourn to Fleming Saunders.[30] Apparently the demand for the company's products was not enough to support the business, however, for on 5 January 1818, 100 acres "comprehending the tilt hammer Establishment" were sold to William Dillon.[31]

Dillon operated the tilt hammer on a reduced scale, mostly making or repairing iron wagon parts, until his death about a year after the purchase. When his estate was inventoried in January 1819, it included "1 Anvill old, 1 Vise, 1 Sledge, 1 bellows, 3 pair tongs, 1 pair knippers, 1 pair compasses and Strainer, 1 Crow bar, . . . 300 lbs. Iron bands, Gudgeons and Arms."[32]

The fact that the tilt hammer business was unsuccessful did not deter the Saunders brothers from their plan to acquire an ironworks. Peter Saunders approached the heirs of Jubal Early and found they were willing to part with their one-sixth share in the Washington Iron Works. On 1 January 1818, the sons and widow of Jubal Early sold their share to Peter Saunders for $13,000.[33] The Saunders brothers next pooled their resources to purchase the interest in the ironworks controlled by the executors of James Callaway. On 5 August 1822, James Steptoe, Christopher Clark, John Callaway, and George Callaway sold their two-thirds share in the Washington Iron Works and their sole interest in the Carron Iron Works to Peter, Fleming, and Samuel Saunders.[34] The price was

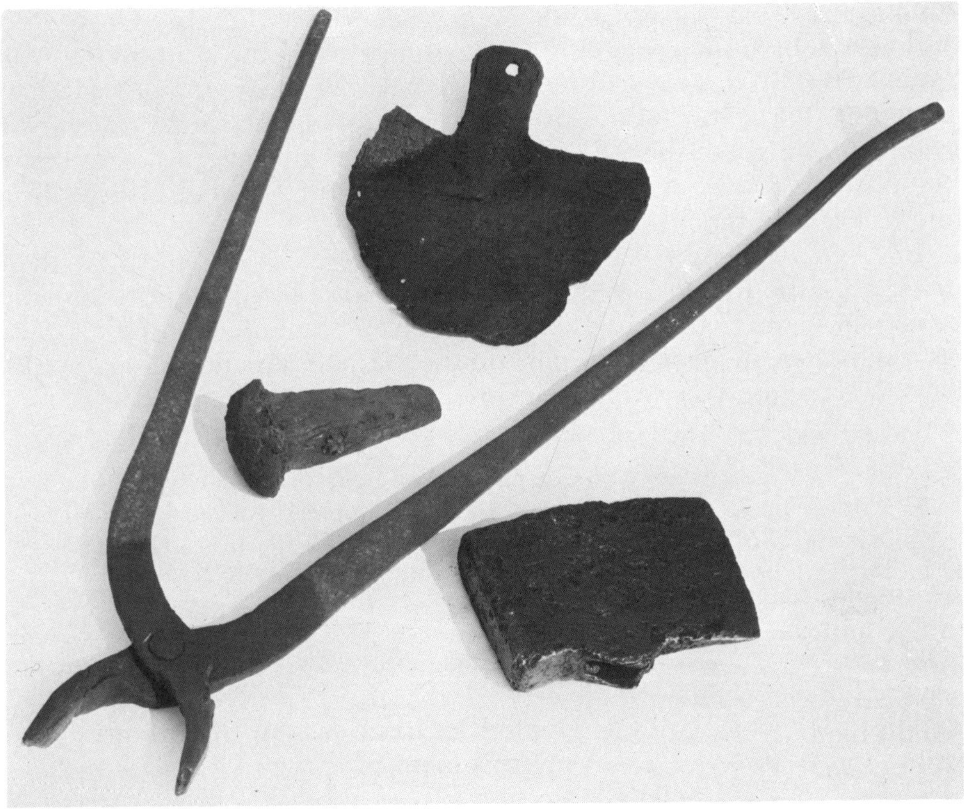

Antebellum ironworkers' tools recovered from the vicinity of the work arch at the Washington furnace include large tongs used to handle hot metal, a small shovel *(top)* that probably was used to dig channels in the sand-casting floor, a wedge *(middle)* for breaking chunks of ore, and *(bottom)* an early axhead.

$30,000. A year later, on 18 September 1823, the three Saunders brothers bought the remaining one-sixth interest in the Washington Iron Works from John Early's son James Callaway Early and his mother, Elizabeth Early, for $20,000.[35] The ironworks were now under the complete control of the Saunders brothers, Peter owning four-ninths of the business in his own right, with Samuel and Fleming owning five-ninths between them.

In 1820, while the Saunders brothers were in the midst of their purchases, the fourth United States census was taken, and it included a survey of manufacturing establishments. Peter Saunders, in his capacity as ironmaster, furnished a report on the Washington and Carron works. The Washington Iron Works consisted of

> one Furnace worked with one pair of Tub Bellows and carried by one water Wheel. . . . One Forge with two Hammers worked by Water Wheels, also Four Fireplaces worked with Water Blast. . . . One Tub Mill with one pair Stones and one Sawmill.[36]

The Carron Iron Works was identical, but was reported inoperative except for the sawmill.

The tub bellows, widely adopted by ironworks in this country after 1800, was considered a vast improvement over the old leather bellows. The new all-wood bellows consisted of two tubs or cylinders mounted side by side on a wooden frame, with a third tub positioned on top. Inside the lower two tubs pistons driven by the waterwheel forced air into the third tub, from which it was conducted by iron, wood, or reinforced cloth pipes to the tuyere of the furnace. This arrangement permitted easier maintenance than did the older variety, and it also provided a more even and more easily regulated blast.[37]

A tub mill was simply a gristmill powered by a tub wheel. In contrast to an overshot or undershot waterwheel, a tub wheel was placed in a position horizontal to the ground rather than vertical. It was made of a single piece of wood if one large enough could be found, and measured about three feet in diameter and almost a foot thick. Blades were chiseled from the wood so that the wheel resembled a modern turbine wheel, and water fell on top of the wheel from a chute, turning the grindstone.[38]

Peter Saunders also furnished information concerning the production of the Washington Iron Works. One hundred and fifty tons of bar iron and hollow ware valued at $120 per ton were manufactured annually, of which almost 142 tons were sold for a gross income of $17,000. Expenses totaled $10,000, so the annual profit was $7,000. The capital investment in the property was estimated at $100,000, which meant that the annual rate of return on the investment was 7 percent.

Antebellum American iron manufacturers generally used tub bellows, or wooden cylinder bellows, in which two pistons *(d)* pumped air to the blast furnace through the sheet-iron pipe at the right *(c)*.

Products manufactured at the Washington furnace and recovered from its site include *(top, from left)* a cannonball, clock weight, fragment of a spice mortar, and piece of wagon hardware; *(middle)* a clevis ring, unidentified piece, malformed cannonball, horse-tethering block, unidentified piece, and cast lap ring; and *(bottom)* a wedge, bracket, unidentified piece, and flatiron.

In addition, Saunders reported on other aspects of the ironworks property. He wrote that

> attached to the above establishment there is Twenty five or Thirty Thousand acres of Land well timbered, and an inexhaustible ore Bank, and elegant sites for any Kind of water machinery.[39]

The 1820 population schedule supplements the information provided by the industry census for that year. At the ironworks were 89 male and female slaves of all ages, of whom 54 were engaged in iron production and 8 in agriculture.[40] The tax books for 1820 list 30 horses in addition to the slaves.[41]

Few of the slaves were specialized or highly skilled workers. Most of them undoubtedly were miners, woodcutters, and wagoners. They were shifted from one job to another as the occasion demanded. When harvest time arrived everyone helped, and if the ironworks ran out of charcoal all hands fell to with axes.

Most of the iron produced at the furnace was hammered into bar iron at the forge, but a considerable number of castings and hollow ware were made as well. Founders poured molten iron into pottery molds through holes called sprues, and when the molds were opened after the iron had cooled the residue found in the sprue was broken off the casting and thrown away.[42] Hollow ware cast at the furnace included pots, pans, ovens, skillets, kettles, and similar kitchen utensils. Other ironware included five- and ten-plate stoves, firebacks, and machinery parts, all produced by pressing carved wooden forms into the sand of the casting bed, then pouring molten iron into the depression left after the form was removed.[43]

The prices of various types of iron products in 1820 can be computed from data compiled in nearby Botetourt County. Pig iron was valued at $30 per ton, castings and hollow ware at $80 per ton, and bar iron at $120 per ton.[44] Pig iron seldom was sold away from the furnace, except perhaps to small independent forges. Cast iron and bar iron were the usual products sold to the public.

Some of the ironworks slaves drove wagons full of iron to landings on the Pigg and Staunton rivers for transportation by shallow-draft flatboat to distant markets, a development that fulfilled James Callaway's dream of shipping iron downriver. An advertisement in a North Carolina newspaper, the *Milton Gazette and Roanoke Advertiser*, by Samuel Watkins and Company on 31 July 1830, informed its readers, "Have on hand and will continue to receive from the Washington Iron Works a complete assortment of Iron, which they will sell at $5 per 100 pounds Cash."[45]

In 1832 products from the Washington Iron Works were being sold in Clarksville, Virginia, also along the Roanoke River. The Saunders brothers' commission agent there, Caleb H. Turner, wrote to Fleming Saunders on 25 April 1832 to inform him that he had sold 6,861 pounds of iron at between $120 and $125 per ton. He added that the market demanded smaller sizes of iron than he had been sent:

> Could have Sold all if it had bin well assorted. The Custum wants light Carry all tire [thin iron bands to fit around the wooden wheels of light wagons]. Have 70 peces of Iron on hand and that Large Iron. Want you to Send two Tuns of Carry all tire and Small bars for horse Shoes and wide bars for axes and [hoes].[46]

Turner wrote again on 7 November 1832, asking for more favorable terms for his purchases and for the brothers' indulgence in allowing him more time to pay what he owed them. "The Iron is all sold," he wrote,

> but about 1000 lb. (it is all large Bars and Sells Slow). Have not Recvd. more than $100 in hand for all the Iron Sold. Have Sold all of it to Custumers to my Blacksmith Shop last year and this. Being Custermers can not push them for the money. . . . And if you would befriend me by indulgen me a little I Could help the Cause very much for I think our market in one or two years will Excell Danvill or any other in this upper Cuntroy.[47]

The Saunders brothers, however, were not impressed with Turner's pleas for more time and decided to terminate their association with him. Turner wrote to Fleming Saunders on 7 December 1832, enclosing his account for settlement. He listed the accounts due Peter Saunders and Company from those to whom he had sold iron, deducting 10 percent for his commission, and including his expenses "for Boating Iron . . . for Halling from the River . . . [and for] 4 peaces of Iron not Delivered per the Boatmen."[48] Turner concluded,

> The Ballance on hand is thirty Bars which I Shall Sel for Cash only or Turn it over to Williams and Co. If you Should make them your Agent here after. If so please to Direct me, other ways I will make Sale of it and Close the Business as quickly as possable.[49]

Five years later the ironworks was selling its iron in Danville, Virginia. On 8 February 1837, Rawlins, Coleman, and Company wrote to Fleming Saunders in Rocky Mount concerning their account with Peter Saunders and Company:

> If you see them shortly please say to them that we have only four or five pieces Iron remaining on hand and that there is none in this market that will Sell, that we have orders and applications for at least

> 3 Tons which we Cannot furnish for want of it, and request them to send down as early as possible with all descriptions of Iron.[50]

Iron may have been transported to nearby places such as Danville by five-horse wagon teams as well as by water. The tradition in Franklin County is that iron was hauled by such teams "to the Carolinas and Georgia, and lumber from these states was hauled back to finish the finest homes in the County."[51] Although it may seem unlikely that it was necessary to have imported lumber from such a distance, it also is true that the demands of the ironworks for charcoal had resulted in the stripping of timber from large areas of Franklin County.

Within a few years of their purchase, then, the Saunders brothers were operating a prosperous business with several good markets for its wares. In addition, they had become respected leaders of their Franklin County community: Fleming Saunders, a prominent lawyer and jurist; Samuel Saunders, a noted planter and landowner; and Peter Saunders, the principal owner and manager of the ironworks. For Peter Saunders, however, respectability and success in business were not enough. He also wanted to dominate the political structure of Franklin County by securing its highest appointive office, that of sheriff.

On 1 November 1830 the county court recommended Peter Saunders to the governor for appointment to the position of sheriff.[52] Again, as in 1808, there were strenuous objections, for there were those who believed that Saunders had disqualified himself as a county justice (and therefore as a sheriff) by his service as a federal postmaster from 1810 to 1815. Anthony Street, who thought he should have been recommended for sheriff over Saunders, wrote to Governor John B. Floyd on 1 November 1830 complaining of the dominance of the Saunders family in local affairs:

> There is such a train of Family Connection to Combat with, I stand alone in the County. There is not one drop of my blood that Circulates through the Veins of any magistrate in the County except my own.[53]

Street elaborated upon this theme in his formal petition:

> Your memo[rialist] begs leave to state there is but verry little Chance for him or any other person to obtain the recommendation of Franklin County Court unless he be a favoured connection or Friend of the Saunders party. I will say without fear of Contradiction that we have four Colonels in our County, three of them are Nephews of Peter Saunders. The Commissioners of the revenue are each of them his Nephews and there is not a youth of the Hairston, Saunders or Hale family that has arived to the age of 20 that is not an officer. The

> deputy Sheriffs, one of them is said Saunders Nephew and the other his brother-in-law, and the Judge of the Superior Court of Law is his brother, and I do assure [you] we are too much under their government.[54]

Like James Callaway before him, Peter Saunders found that his prosperity resulting from the ironworks business was enough to earn him the emnity of certain people, Anthony Street in particular. Street complained to the governor that he lived

> say 16 or 18 miles from our Courthouse and have done he believes more of the drudgery of the office of Justice of the peace than any man in the County. My opponent lives in 5 or 6 hundred yards of the Courthouse. . . . He is a wealthy man and one of wealthy Connection and many of them in this County and some in other Countys. Those things are advantages I possess none of. I have no connection in this County, only my own family.[55]

Anthony Street succeeded in his efforts to deny Peter Saunders the sheriff's office for several years. It was not until 10 December 1835 that Saunders finally received a commission as sheriff.[56] A friend of the family wrote to Fleming Saunders on 20 May 1836, "I was very glad to hear that Colonel Peter was Commissioned Sheriff. I have no doubt it was gratifying to him on many accounts."[57] Peter Saunders was recommended for a second term as sheriff on 7 November 1836, and was again appointed.[58]

Peter Saunders was now at the peak of his career: wealthy, influential, and held high enough in the esteem of his peers to serve in the most honored office in the county. The words of one historian of the iron industry apply almost as well to Saunders as to the Pennsylvania ironmasters, who the author said

> were almost feudal lords, to whom their workmen and their workmen's families looked for counsel and guidance in all the affairs of life as well as for employment; whose word was law; who often literally owned their black laborers . . .; who cultivated farms as well as made iron; who controlled the politics and largely maintained the churches and schools of their several neighborhoods; who were captains and colonels of military organizations; . . . whose dwellings were usually substantial structures, which were well furnished for that day and ordered in a style of liberality and hospitality.[59]

6

The Iron Plantation

The southern iron industry reached its economic peak about the same time that Peter Saunders attained his pinnacle of personal achievement. The price of iron was rising yearly, and the cheap labor furnished by the slaves increased profits. Slavery, long considered primarily an agricultural institution, had been adapted over the years to manufacturing and industry with little difficulty. Slaves labored at almost every task imaginable on the iron plantation.

> At the furnaces and forges slaves tended fires, worked the metal, and in fact did everything but manage the establishments, which was always the job of a white man. Elsewhere on an ironworks plantation Negroes planted and harvested crops, cut and charcoaled wood, mined iron ore, drove wagons and manned boats, made shoes, ground flour, and worked as carpenters and blacksmiths. The elite among them were the refiners, molders, and blacksmiths, and because a skilled slave was as valuable as two ordinary hands, many ironmasters owned a few skilled workers.[1]

The type of work assigned to a slave varied depending on the season and the requirements of the ironworks. When the furnace was in blast the slaves worked in two twelve-hour shifts, dumping ore, flux, and charcoal into the furnace, opening the taphole so the iron could flow into the casting bed, hammering out bar iron, tending the charcoal heaps, and hauling ore. The planting and harvest seasons commanded the labor of all hands available, and as on any plantation there were dozens of jobs aside from agriculture and iron founding that needed to be done.

> It was common practice on plantations to allow slaves to do extra work for wages, and this was also permitted at ironworks, where Negroes did such labor as driving wagons, chopping wood, and making

> baskets, tar, spikes, and rails. Extra work could also be done at night, on Sundays, and over the Christmas holidays. This labor was credited to a slave's account, and . . . the Negroes usually spent their money at ironworks' stores, where they bought articles of all kinds, especially clothing.[2]

The treatment of the ironworks slaves, as with agricultural slaves, varied from one place to another. In some places food was distributed among the slaves who then cooked it individually; in others, the food was cooked in a central kitchen and was eaten by the slaves in groups, so they could be better supervised.[3] Discipline was often harsh: "In Rockbridge County a Negro is said to have hurled himself into a furnace rather than face a whipping."[4] Regardless of the severity of the discipline, the ironmaster often was faced with the problem of runaways.

The Washington Iron Works was typical of the southern iron industry and had grown into a great iron plantation. While many southern ironmasters hired slaves to work for them, the Saunders brothers owned the slaves at the ironworks. The number of slaves over twelve years of age varied from 64 in 1822, when the brothers obtained full control of the Washington Iron Works, to 83 in 1837.[5]

Most of the slaves at the Washington Iron Works lived in cabins on Scuffling Hill across Furnace Creek from the blast furnace. The hill is said to have derived its name from the fact that it was a "long hard scuffle" up its steep slope after a twelve-hour day's labor at the ironworks.[6] The slaves tended vegetables and other crops grown near Peter Saunders's house, convenient to the kitchen, and over the years the house and its grounds became known as the Farm at the Furnace.[7]

James Callaway's old residence and ordinary at the ironworks had been enlarged by Peter Saunders from a story and a half to two full stories. The short planks that were added to the floor on the second story to widen it are still in place. Saunders family tradition alleges that the southwest room on the second floor was "Uncle Peter's room."[8] Jane Jones Saunders, who acted as hostess for her father, presumably occupied the other upstairs room. Near the house stood various plantation outbuildings, including an office, a dairy, a barn, the stables, a carriage house, a blacksmith's shop, and a two-room brick structure that doubled as a kitchen and as a dwelling for the house servants.

Peter Saunders's gracious lifestyle, of course, depended upon the prosperity of the ironworks business. In 1836 Joseph Martin in his *Gazetteer of Virginia* described the ironworks as a thriving enterprise. "In the vicinity [of Rocky Mount]," he wrote, "there is an iron furnace and forge, which give employment to 100 operatives, and manufacture about 160 tons of bar iron and castings annually."[9] Martin also noted that the iron was "of a very superior quality."[10]

Ironware, some manufactured at the Washington Iron Works, displayed in the brick kitchen behind the ironmaster's house. The biscuit beater on the butcher block and the andirons are probably local products. A spice mortar rests on the shelf to the left, muffin pans lean against either side of the fireplace, and a variety of pothooks, lid lifters, spoons, and suspension hooks hang at the right. A griddle and large kettle (similar to that on page 40) hang from the original crane, while the hearth displays *(from left)* a small pot, coffee roaster, dutch oven, rotating broiler, dutch oven without its lid, griddle or skillet, long-handled toaster, trivet, and large pot.

The quality of the iron was due in part to the high grade of iron ore used at the furnace. At the end of July 1839 George W. Boyd, an assistant state geologist with the state geological survey conducted by Professor William Barton Rogers at the University of Virginia, arrived in Franklin County while on a geological tour of the southwestern piedmont region of Virginia. On 2 August Boyd explored the vicinity of the Washington Iron Works and recorded his observations in his field notebook:

> In the morning rode . . . West and Southwest from town. About 1 mile Southwest from Court House . . . is the principal ore bank which supplies the Furnace at this place. It is a bed or layer of Magnetic oxide of Iron, granular, color black and greenish black, associated with decomposing micaceous slates. . . . The black has the purest look, but is not so, for it contains much Sulphur or rather Iron pyrites and hence this sort is not used. . . . West of the ore bank on waters of Furnace Creek . . . are large masses of a hard siliceous rock, which is Greenstone or an altered Granitic rock [used to construct the furnace].[11]

On 5 August 1840 Boyd left Rocky Mount and rode eighteen miles east to Anthony Brooks's ordinary to visit the only large deposit of limestone known in the county. The next day Boyd left Brooks's and took

> the road to ford of Pig River. Near the ford turn Northeast along a small branch to visit the Limestone, which is about 1/2 mile from the road. It is exposed on this small run, and is quarried for the Furnace at Rocky Mount, where it is used as a flux for the Iron ores, and it is also burned into Lime by Mr. Rives to whom the land belongs.[12]

Forty-three years later, in 1882, another geologist visited Franklin County. Professor William M. Fontaine of the University of Virginia was given an account of the Washington furnace's operation at the peak of its prosperity. The furnace, he wrote,

> had the reputation of making excellent iron. It was highly valued for all the purposes for which wrought-iron is used. . . . It was always a cold blast, charcoal furnace, and made about 4 tons of iron per day. Only the outcrop of the ore, for about 20 feet down, was used, as this only gave the most easily managed iron, with the appliances then used. . . . The thickness of the deposit ranged from 2 to 4 feet.[13]

Despite the quality of the ore, the security and prosperity of the iron plantation were almost destroyed by the Panic of 1837 and the years of industrial depression that followed it. The 1830s had been boom years for the charcoal iron industry, which had seen the price of pig iron rise from $35 per ton in 1830 to $52.50 per ton in February 1837, an increase in price of two-thirds. Then in March the price began to fall, and by October it had dropped to $34.75, less than at the beginning of the decade, reflecting the lack of demand caused by the panic. The iron industry sank further into depression over the next several years as the price of pig iron continued its fall to a low of $25 in the summer of 1843.[14]

A variety of financial disasters produced the Panic of 1837 and the industrial depression that followed it. Chief among them were a chronic shortage of hard currency, the calling in of loans by American and British banks, inflation, and the failure of over six hundred banks in the United States. Unemployment rose while industrial production declined. The southern iron industry was struck both by the panic and by superior northern manufacturing techniques.

The feature of the southern iron plantation of the antebellum period that most distinguished it from its northern counterpart was the institution of Negro slavery. The southern ironmaster was assured of a stable source of relatively cheap labor, but he was required to support those slaves too young, old, or sick to work along with those who were capable of labor. There was another, less obvious disadvantage to slavery. The northern ironmaster, because he had to pay more for workers' salaries, was compelled to find other ways of reducing his overhead and making his operation more efficient. Experimentation in the North showed that coke and anthracite coal were more economical fuels than charcoal, and the puddling furnace, in which the iron was stirred as it melted, was found to be a technological improvement over the older style of blast furnace. Because the South had lacked the North's incentive for innovation the pace of change had been slow, and by the 1840s northern iron could be purchased in the South at a lower cost than southern iron.[15] The southern charcoal iron industry had begun to lose its competition with the northern iron industry long before the Civil War, and the Washington Iron Works was affected by this loss as well.

Peter Saunders was forced to cut back production at the ironworks and reduce his expenses by every means available. One step he took was to dispose of most of his slaves by selling or renting them in 1837 to his nephews Giles Hale and John S. Hale, who then assumed the responsibility for their care and feeding, as well as for the taxes due on them. In 1840 Saunders recovered his slaves in hopes that the industry soon would rebound.[16]

Over the next two years Peter Saunders and his brothers sought a way to save the ironworks by dividing the land and slaves among them to spread the tax burden more evenly. In 1840 Peter Saunders was allotted 700 acres of land and the house in which he lived for his own use. In 1842 there was, he wrote,

> a division of the Iron works Lands and Negroes and my part has been allotted to me and a Deed in part executed for the Lands and Negroes . . . to my proper use and the tract of land I live upon has been increased to one thousand Acres.[17]

The brothers had decided that to render their property more secure, and for the better management of the ironworks, Peter Saunders would be given full control of the land, while Samuel and Fleming Saunders would take over the ownership of the slaves. On 7 January 1842 Samuel and Fleming Saunders and their wives conveyed to Peter Saunders their interest in 6,000 acres deeded to the brothers by James Callaway's executors in 1822.[18] On the same day Peter Saunders made a deed of the same land to his brothers, probably so that if he died suddenly Samuel and Fleming Saunders could present the deed for recording at once and take full legal control of the ironworks.[19] For the time being, however, Peter Saunders's brothers merely filed his deed away for safekeeping.

On 17 September 1842 Fleming and Samuel Saunders and their wives sold to Peter Saunders for one dollar all their interest in the Furnace Creek tract of 1,000 acres that earlier had been allocated to him.[20] This completed the transfer of property that was now reflected in the tax lists. The slaves who had been listed under "Peter Saunders and Company" in the personal property tax books were, beginning in 1842, listed under "Samuel and Fleming Saunders."[21]

Although the brothers' plan kept the ironworks in their possession, the depression in the iron industry continued unabated. Perhaps it was the depression that reminded Peter Saunders of his own mortality. On 4 November 1840 he wrote his will, in which he left to his "natural daughter Jane J. Saunders who now resides with me all the interest both real and personal which I own in the property known as the Iron works property."[22] He also left to her and her descendants, if any, a number of slaves and a house and lot in Mount Pleasant adjoining the town of Rocky Mount. If she died without issue the property was to go to Fleming and Samuel Saunders. Peter Saunders added that it was his wish

> that the property known as the Iron works property be kept together so long as my daughter Jane aforesaid and my Brothers Fleming and Samuel Saunders shall find it in their interest to do so.[23]

Peter Saunders had ambivalent feelings about the institution of slavery and the hardships imposed by law upon free Negroes. Although he was willing to rely on slave labor for his own support, Saunders also showed unusual concern for the welfare of five Negroes who had served him. "I hereby declare," he wrote,

> that the Negroe woman Nancy and her children Anne, Pamela, Harriet, and Alfred are free by Birth, and it is my desire and bequest

> that they shall enjoy their Freedom. But should any difficulty arise with regard to their remaining in the state of Virginia I hereby declare and bequeath them to my Daughter Jane aforesaid to hold for her life time as a guardian for their benefit and advantage with the power of appointing by will another trustee and guardian for them, provided that they shall never be subjected to servitude.[24]

He left the remainder of his slaves to Jane Jones Saunders with the provision that they be "hired out and the proceeds thereof applied to the support of said Nancy and her children." Saunders also directed that two tracts of land be sold for the same purpose. In a codicil to his will, written on the same day, Peter Saunders bequeathed "the House in which I live together with seven Hundred Acres of Land more or less lying around it" to his daughter Jane.

Four years later, Peter Saunders retired from the ironworks. On 2 November 1846 he sold the Furnace Creek tract to Samuel and Fleming Saunders,[25] and on 14 November he purchased an 800-acre plantation in Pittsylvania County named Belle Grove from Whitmel P. Tunstall.[26] Two days before the deed to Belle Grove was executed, Tunstall and Saunders signed an agreement concerning the sale. Peter Saunders agreed to give Tunstall 400 acres of land on Blackwater River in Franklin County and two other tracts near Rocky Mount, one of 37 acres and the other of 40 acres, "with a negroe man by the name of Gabriel, together with obligations on Mssrs. Fleming and Samuel Saunders for the sum of six thousand dollars."[27] Tunstall agreed to deliver to Saunders, together with the Belle Grove tract,

> 150 Barrels of corn, 10,000 lb. of Oats, the Top stacks, straw stacks, all the farming utensils such as ploughs, a two Horse wagon, the stock of Hogs and cattle, except Killing or fattening Hogs and two sows, together with two yokes of steers with cart and the cattle except two Milch cows and calves and two Heifers.

Peter Saunders, meanwhile, had written a new will on 7 February 1843, leaving his estate to his daughter. He also provided that if she died without issue then all of his slaves were

> to be emancipated and if not permitted to remain in the commonwealth of Virginia, then my Lands and other remaining estate in her hands at her death to be sold and the proceeds arising therefrom to be invested in Lands, Farming Tools, and provisions for their comfortable settlement in some one of the states of the union as may permit them to settle therein and said Lands to belong to them in Fee [simple].[28]

Because of his purchase of Belle Grove, Peter Saunders added a codicil to his will on 15 April 1847 specifically leaving the property to his daughter. Thirty-six days later he was dead.

Saunders's retirement after almost thirty-five years at the Washington Iron Works had been brief and beset with illness. Soon after he moved to Belle Grove he suffered a paralytic stroke that greatly incapacitated him. On 11 April 1847 he had a second stroke, which was described by Louisa Saunders in a letter to her father Fleming Saunders:

> Cousin Jane has requested me to write you my Dear Father to inform you that uncle Peter has had a second atact [*sic*] of paralicis since I last wrote home. All this week he has had a sudden prostration of all his powers about 8 or 9 O'clock in the morning. The fit was more violent this morning but he soon recovered the use of his limbs again and is now better. I don't know when I was so shoocked [*sic*] as I was this morning when I beheld my uncle born aparantly lifeless in the arms of four men. Poor Cousin Jane is in a state of great distress. Doctor Reeves was here this morning [and] says it is an attack of paralicis withought doubt.[29]

Although Peter Saunders rallied enough to add the codicil to his will on 15 April, the strokes ultimately proved fatal. He died on 25 May 1847[30] and was buried in the Tunstall graveyard behind Belle Grove. On 7 June 1847 the *Lynchburg Virginian* printed an obituary:

> Died, at his residence, in Pittsylvania county, Col. Peter Saunders, in the 72nd year of his age. The deceased was a native of the county of Franklin, where he resided until within a few months of his death. Correct and honorable in his intercourse with mankind, kind in his feelings, courteous in his deportment, and hospitable to all, he lived to share largely in the affections of his fellow citizens and died sincerely lamented.[31]

Why Peter Saunders moved from the county of his birth to Pittsylvania County has never been explained. Family tradition claims that, whatever the reason, the old man soon changed his mind and sought to return to his house at the Washington Iron Works.

> In his last illness he begged to go back to die. His family realized he was too sick to move but the negros being very superstitious thought he should be carried back and were sure he would come back to haunt the house. Even to this day the room he used to occupy is known as 'Uncle Peter's Room,' and when a strange noise is heard, such as is frequently heard in an old house, someone in the family will remark, 'Uncle Peter is paying us a visit.' Strange to relate the noise usually sounds as if someone were walking up the front stairway to his old room.[32]

7

The Last Years of the Washington Iron Works

After the retirement and death of Peter Saunders the Washington Iron Works came under the full control of Fleming and Samuel Saunders, neither of whom had any disposition to manage the ironworks personally. Fleming Saunders was sixty-eight years old in 1846, a jurist, farmer, and resident of Campbell County. His brother Samuel Saunders, who lived in Franklin County, was sixty-three years old and almost fully occupied with his huge plantation. Since they had left the daily management of the ironworks in the hands of Peter Saunders, neither brother had the experience necessary to operate it.

The brothers decided that Samuel's son Peter, who probably had learned the craft under his uncle Peter Saunders, would take over as ironmaster. The new ironmaster, locally known as Captain Peter Saunders because of his militia rank, was born about 1815.[1] Fleming Saunders also had a son named Peter. He was born on 6 October 1823,[2] became a lawyer and politician, and resided at the old family home at Bleak Hill. To distinguish between the two cousins, Samuel's son was known as Captain Peter Saunders or Peter Saunders, Sr., while Fleming's son was called Peter Saunders, Jr.

The first order of business for Captain Peter Saunders was to settle the claims of Jane Jones Saunders against the ironworks. A careful accounting for the years from 1842 to 1848 showed that there was a balance due her father of $91.94, and on 20 January 1848, that sum was "Transferred to Credit of Miss Jane J. Saunders in full."[3]

On 1 January 1849 Fleming Saunders sold his share in the old Carron Iron Works property to Samuel Saunders.[4] Their brother Peter had sold his share to the two of them on 7 January 1842,[5] and with Fleming's sale, Samuel Saunders became sole owner. On 31 August 1849 Samuel Saunders sold the tract to two of his sons, Captain Peter Saunders and Robert Hyde Saunders.[6] Before long they began to restore and rebuild the Carron Iron Works.

In the meantime, however, the Washington Iron Works occupied most of the time and energy of Captain Peter Saunders. The business was scarcely a money-making enterprise any longer due to the prolonged depression of the iron industry and the increased competition from northern manufacturers. Saunders deemed it necessary to go into debt for supplies that the plantation could not produce, but this hardly pleased his father and uncle since they depended upon the ironworks for part of their income and felt that it reflected little credit upon his managerial skills. Writing to his uncle Fleming Saunders on 3 March 1848, Peter Saunders commented upon his financial difficulties:

> I am not able to say whether I shall be able to spare you any funds from the Iron works or not. If I meet the Tunstall's debt promptly which I am determined to do I fear it will take all the funds that I can conveniently raise. I shall however do all I can. It would be my pleasure to have money for you and my father to pay your debts and to spend. You must recollect that I have been hard run ever since I have been at the Works, having to expend a great deal upon the improvement of the property and one or two large debts to pay. The debt to uncle Peter of something upwards of $1,100.00 and one to John S. Hale of $600.00. Those debts are now very nearly paid but a small balance on each.[7]

Among the improvements planned by Captain Peter Saunders was the purchase and installation of a new type of waterwheel invented by Isaac L. Dripps, superintendent of machinery for the Camden and Amboy Railroad, in New Jersey. Saunders wrote to Joseph R. Anderson, head of the Tredegar Iron Works in Richmond, for his opinion, having heard that Anderson had had some experience with the wheels. Anderson replied on 29 May 1848, but his attitude toward Mr. Dripps and his wheels was anything but favorable:

> The Armory Iron Company in which I am interested put up two of Dripps Wheels about 12 Months ago and the best opinion I can give you about it, is to say that after having given it a thorough trial under a great cost . . . we are now stopped and are removing them and putting in an overshot wheel. . . . Altogether they have cost us directly and by losses they have occasioned from Ten to Fifteen thousand Dollars. . . . From what I have seen I would not use it for any purpose if the Patentee would make a present of it and put up and attach it gratis. I think the old kind of wheels are after all the best altho you had better make some allowance for my having been taken in and suffered much by the experiment.[8]

Although Saunders saved money by following Anderson's advice, he increased his indebtedness by expanding his operations to produce more iron. After he and Robert Hyde Saunders purchased the Carron Iron Works property in 1849, the two brothers immediately set about rebuilding the old works. They completed the Valley Forge on Pigg River early in 1850 and soon had it in operation. According to the 1850 census the Valley Forge had produced 70 tons of bar iron valued at $5,600, or $80 per ton. The brothers had a capital investment in it valued at $13,000, and employed twelve hands at a total monthly expense of $240.[9] The forge had "2 refinery fires, 1 run-out and 1 hammer worked by water," and during the next decade it produced about 150 tons of bar iron annually.[10]

Despite its financial difficulties the Washington Iron Works also was in production in 1850. According to the census, the total capital investment in both the furnace and the forge was $80,000. Fourteen men and six women worked at the furnace, and another fourteen men worked at the forge. The furnace produced 300 tons of pig iron annually, valued at $7,500, or $25 per ton, while the annual production of the forge was 160 tons of bar iron valued at $12,800, or $80 per ton.[11]

Captain Peter Saunders may have hired several white laborers to work at the ironworks in addition to the slaves. According to the census of 1850, a number of his near neighbors, who may have been living in dwellings on the plantation, were laborers: James Doughton, aged thirty-two, John C. Tyre, aged thirty; Elijah Bird, aged sixty-five; and his son Joseph Bird, thirty-two. James Blair, aged sixty, listed on the census as an "Iron founder," lived in the household of John C. Tyre.[12]

It was in 1850, according to local tradition, when the Washington Iron Works ceased operations forever, except for a brief period during the Civil War. The earthen dam on Furnace Creek burst during a heavy rain, and the water stored to power the blast machinery flooded the furnace while it was still hot. The superheated rocks lining the boshes exploded, cracking the exterior of the west wall of the furnace from top to bottom.[13] One source relates that "there is an old story in the Saunders family that the explosion was so remarkable that a notorious horse thief in hiding nearby was so frightened he came out and gave himself up to the law."[14]

The ruined furnace became a drain on the Saunders family finances. Fleming Saunders wrote to his sons in Franklin County on 30 January 1852 to encourage them to settle the affairs of the ironworks. "Press your Cousin Peter, etc.," he wrote, "to settle their Ironworks accounts, and if practicable to send me some money as I am almost entirely without."[15]

Captain Peter Saunders, however, was extremely reluctant to wrap up the business, perhaps because he hoped to rebuild. He had other projects under way, but his kinsmen quickly wearied of waiting for him to act. In a letter to his father, Fleming Saunders, on 3 March 1852, Peter Saunders, Jr., complained of his cousin's management:

The Washington furnace photographed from the southeast, showing the work arch and casting floor in the foreground and a tuyere arch at the right. A bridge ran from the furnace to the hill behind and an earthen dam to the left channeled the waters of Furnace Creek along sluices to the waterwheels that drove the tub bellows behind the furnace.

> I don't know what is to be done with the Iron Works. It is making nothing in its present condition. The Captain is very dilatory about making an exhibition of the state of the accounts. He has entirely too many Irons in the fire and under its present management the property will only be involved in debt. I don't like to find fault but the property is not being managed as it should be. It would perhaps stimulate the Captain were you to write him and instruct him positively to make as soon as possible an exhibition of the state of his accounts. Until that is ascertained it is impossible to advise what disposition should be made of the property.[16]

One factor that rendered an immediate accounting necessary was the death of Samuel Saunders during the preceding winter. On 5 January 1852 the administration of his estate was granted by the Franklin County court to two of his sons, Peter Saunders, Jr., and John Ingles Saunders, and to his son-in-law John S. Hale.[17] Strangely enough for a man who had served as a justice on the county court for most of his adult life, Samuel Saunders died without a will. The size of his estate and the large number of his heirs complicated the division of his property, and, in addition, the administrators were faced with the task of making an equitable division of the ironworks property owned jointly by Samuel and Fleming Saunders.

On 6 December 1852 the matter was presented to the Franklin County Court for adjudication, and a division of the ironworks lands was ordered.[18] The division soon was completed to the satisfaction of all the parties, and on 4 January 1853 the settlement was recorded.[19] The land owned in partnership by Fleming and Samuel Saunders totaled 15,282 acres lying on both sides of Rocky Mount. Conveniently, it was split almost in half by the Pittsylvania, Franklin, and Botetourt Turnpike, which followed approximately the route now taken by Main Street from Pigg River northward. The land to the east of the turnpike, which totaled 7,542 acres, was assigned to Fleming Saunders, and the land to the west, 7,710 acres, was distributed among the heirs of Samuel Saunders. John Ingles Saunders received 522 acres, including the ironmaster's house and the Washington furnace.

Three years after he took possession of the property, John Ingles Saunders remodeled the house, removing the last vestiges of its colonial interior.[20] It probably was at this time that the two-story wing was attached to the rear of the house making it T-shaped. The remodeling of the building was so extensive that Saunders and his wife, Ann Booth Saunders, were compelled to live in a cabin while the work was under way.[21] When the job was completed the house stood essentially as it does today, except that the front porch was only one story high,[22] and there were only two windows on the first floor front instead of four. The rear addition gave the house two more rooms, making six altogether, and a rear stairway. The closets that flank the fireplaces throughout the house may have been added at this time as well.

Although Captain Peter Saunders had nothing further to do with the Washington Iron Works property, he continued to carry on the ironmaking business by building a new charcoal blast furnace on the site of the old Carron Iron Works on Story Creek about ten miles west of Rocky Mount. The furnace, which he owned in partnership with his brother Robert Hyde Saunders, was named the Carron Furnace. It was "built in 1857, 8 feet across the bosh by 30 feet high,"[23] and was in operation by 1860.

On 23 May 1858 Fleming Saunders died, the last of the three brothers who had bought the Washington Iron Works from the heirs of James Callaway and Jeremiah Early. The *Lynchburg Virginian Tri-Weekly* carried a brief obituary: "Died, at his residence in Campbell County, on Sunday the 23rd instant, Judge Fleming Saunders, in the 80th year of his age."[24] The judge's will, written on 28 November 1856, directed that his lands in Franklin, Henry, and Patrick counties be sold, the money to go to his wife. He also left her his shares of stock in the Roanoke Navigation Company, the Virginia and Tennessee Railroad Company, and several turnpike companies. He appointed his sons Peter, Robert, and Fleming Saunders as executors.[25]

Although the Washington Iron Works stood in ruins at the end of the decade, the Carron Furnace and the Valley Forge were in production. However, both were losing money. The Carron Furnace represented a capital investment of $14,000 in both real estate and personal property by 1860, and fifteen men and five women were employed in its operation at an average cost of $213 per month, or $2,556 per year. The furnace consumed 1,000 tons of ore and fuel annually, valued at $4,000, to produce 200 tons of pig iron and castings valued at $6,000, or $30 per ton. The ore and fuel expense plus the wages resulted in a net annual loss of $556.[26]

The Valley Forge was losing even more money than the Carron Furnace. The total capital investment at the forge was down to $10,000 in 1860 from $13,000 in 1850. It took 150 tons of pig iron valued at $6,000, and 60,000 bushels of charcoal valued at $1,200, to produce 90 tons of bar iron a year for an income of $9,000. In addition to the $7,200 spent annually on pig iron and fuel, fifteen male laborers earned average monthly wages totaling $270, or $3,240 per year. With an annual income of only $9,000, and expenses totaling $10,440, there was a net annual loss of $1,440.[27]

Another of Samuel Saunders's sons, Crockett Ingles Saunders, was also involved in making products from iron, and he was able to operate at a profit. Although no record has been found among the Franklin County Court records, it seems that John I. Saunders allowed his younger brother Crockett to salvage foundry equipment from the Washington Iron Works and set up a small plow manufactory. Crockett Saunders operated his enterprise on a tight budget, with a capital investment of only $500, and with four male employees operating the machinery by hand and by waterpower for $50 a month altogether. The raw material used annually, including fuel, was 30 tons of pig iron worth $900. The little shop produced 500 plows a year for a total income of $3,500 and a net profit of $2,000.[28]

Soon the war swept all that away. The freeing of the slaves and the triumph of northern technology doomed the southern charcoal iron industry. A few old furnaces continued to operate in remote corners of the South after the war, but by and large they only supplied small amounts of iron to a local market. The days when southern iron could compete with northern iron were gone. For Franklin County, the Civil War meant the total destruction of the iron industry that had flourished for almost ninety years.

For one brief moment just before the end, however, it seemed as though the war might be the salvation of the several ironmaking enterprises run by Captain Peter Saunders. A local tradition that the Washington Iron Works was somehow patched together to make iron for the Confederacy receives some support from a court order entered on 19 December 1862:

> Peter Saunders Sr. made application to this Court to exempt his slaves from the requisition of the Governor of this Commonwealth for slaves to be used for the public defense, upon the grounds that his slaves are engaged in making iron for Billharts [*sic*], Hall and Co. of Pittsylvania County and John S. Brown of this County to be used by them in making guns for the Confederate States. The said Peter Saunders Sr. proved the said facts to the satisfaction of the Court, but the Court refused to exempt said slaves from said requisition, upon the grounds that this Court has no power under the law to make said exemption and it is ordered that these facts be certified to the Military authorities of the Confederate States at Richmond.[29]

The rejection of Saunders's application probably brought a quick end to his ironmaking activity.

Jane Jones Saunders, who had continued to live at Belle Grove in Pittsylvania County after her father's death, was visiting at the Grove, the home of John S. Hale in Rocky Mount, when she died on 5 December 1861, "in the 63d year of her age."[30] The last principal in the story of the Washington Iron Works was gone.

Notes

Short Titles and Symbols

Duke	Manuscript Department, William R. Perkins Library, Duke University, Durham, N.C.
UVA	Manuscripts Department, University of Virginia Library, Charlottesville.
VHS	Virginia Historical Society, Richmond.
VSL	Virginia State Library, Richmond.
WPA	Works Progress Administration Historical Inventory, General Library Division, Virginia State Library, Richmond.

• • •

Chapter One: Iron Manufacturing in Colonial Virginia

1. William Byrd, *The Prose Works of William Byrd of Westover,* ed. Louis B. Wright (Cambridge, Mass., 1966), 349–354.
2. Ibid., 348.
3. Ibid., 354.
4. *Virginia Gazette* (William Rind), 21 Apr. 1768, p. 2, col. 1.
5. Ibid., 22 Oct. 1772, p. 3, col. 1.
6. *Virginia Gazette* (Alexander Purdie and John Dixon), 10 June 1773, p. 2, col. 2.
7. The history of the Oxford ironworks is unclear due to the nature of the deed of trust. If the owners of the bloomery repaid the debt to James and Robert Donald and Company, they would have kept the property. Otherwise, the Scottish firm would have become the owner of the ironworks and almost certainly would have recorded the deed of trust as proof of ownership. No deed of trust has been found recorded in Bedford County or in Chesterfield County, where Rocky Ridge was located, which may mean that Harper, Elledge, and James were able to repay their debt to the Donald brothers, or that the Scottish merchants neglected to record their deed.

Conflicts between local tradition and available documentation obscure the next change of ownership. Tradition holds that by the time of the revolutionary war the bloomery had been replaced by a blast furnace, and that this furnace was owned and operated by Colonel James Callaway, of Bedford County (T. Gibson Hobbs, Jr., "The Oxford Furnaces in Campbell County, Virginia," in *Lest it be Forgotten: A Scrapbook of Campbell County, Virginia* [Altavista, Va., 1976], 139). It is further believed that about 1779 Callaway sold the works to David Ross, perhaps the wealthiest man in Virginia, who closed the old furnace and built a larger one (ibid., 139).

In contrast to this traditional story, letters survive that were written by Ross in October and November 1776 from the "Oxford Iron Works" (Ross to John Hook, New London [Oct. 1776], Accession 19904, and Ross to Hook, 30 Nov. 1776, Accession 19906, Personal Papers Collection, VSL), indicating that he was in possession of the property at that time. The relevant records of Dinwiddie County have been destroyed, and no deed to the ironworks involving Callaway and Ross has been found in Bedford County records; one cannot determine when the transaction occurred, or indeed whether Callaway ever owned the Oxford ironworks.

Chapter Two: John Donelson and His Bloomery

1. Marquis James, *The Life of Andrew Jackson* (Indianapolis, Ind., 1938), 48.
2. Maud Carter Clement, *The History of Pittsylvania County, Virginia* (Lynchburg, 1929), 155.
3. Ibid., 42.
4. Halifax County, [Court Orders] Pleas No. 3, 1759–1762, Reel 55, p. 222, VSL.
5. Ibid., 340.
6. Halifax County, [Court Orders] Pleas No. 4, 1763–1764, Reel 55, p. 279, VSL.
7. Ibid., 340.
8. Pittsylvania County, Court Records (Orders) No. 1, 1767–1772, Reel 47, p. 1, VSL.
9. Ibid., 21–22.
10. Pittsylvania County [Record Book, Land Records], 1737–1770, Reel 70, p. 126, VSL.
11. Pittsylvania County, Old Surveys, No. 1, 1746–1782, Reel 69, p. 70, VSL.
12. Ibid., 71.
13. Land Office, Patents No. 35, 1762–1764, Reel 35, pp. 532–533, 535, VSL.
14. Land Office, Patents No. 34, 1756–1762, Reel 33–34, pp. 549–550, VSL. This land first had been surveyed for Robert Hill (Halifax County, Surveys, No. 1, 1751–1901, Reel 84, p. 55b, VSL), who had transferred his claim to Wilcox.
15. Joseph Willcox, "Historical Sketches of Some of the Pioneer Catholics of Philadelphia and Vicinity," in *Records of the American Catholic Historical Society of Philadelphia* 15 (1904): 431–432.
16. Wilcox to Joseph and George Morris, Philadelphia, in Morris Family Papers (Bruce Cotten Collection), North Carolina Collection, Wilson Memorial Library, University of North Carolina, Chapel Hill.
17. Willcox, "Historical Sketches," 432.
18. Bushrod Washington, *Reports of Cases Argued and Determined in the Court of Appeals of Virginia* (Richmond, 1798–1799), 1:38.
19. Ibid.
20. Land Office, Patents No. 42, 1773–1774, Reel 41, pp. 750–752, 758–759, VSL.
21. Land Office, Plats and Certificates, 1790–1791, VSL. Upon this consolidated survey a land grant was issued to James Callaway and Company on 12 July 1790.
22. Pittsylvania County, Tithables, 1767–1785, p. 58, "A List of Tithables for Pittsylvania County taken By John Donelson 1773 . . . Tithes at the Iron works," VSL.
23. Michael G. McCollum, "Geologic Occurrence of Iron in Franklin County" (brief typescript report furnished the author by J. Francis Amos, Rocky Mount, Va.), 2.
24. Frederick Overman, *The Manufacture of Iron . . .* (Philadelphia, 1850), 245–247.
25. Ibid., 104–107.
26. Frederick Overman, *A Treatise on Metallurgy . . .* (New York, 1852), 543.
27. Ibid., 545.
28. Ibid., 598–600.
29. Pittsylvania County, Deed Book No. 3, 1772–1774, Reel 1, pp. 359–361, VSL. Deed recorded 23 Sept. 1773.
30. H. R. McIlwaine, Wilmer L. Hall, and Benjamin J. Hillman, eds., *Executive Journals of the Council of Colonial Virginia,* 6 vols. (Richmond, 1925–1966), 6:361.
31. Jack M. Sosin, *The Revolutionary Frontier, 1763–1783* (New York, 1967), 33.
32. James, *Andrew Jackson,* 49.
33. Sosin, *Revolutionary Frontier,* 131.

34. James, *Andrew Jackson,* 49.
35. Henry County, Deed Book No. 1, 1777–1779, Reel 1, pp. 300–304, VSL.
36. Land Office, Plats and Certificates, 1790–1791, VSL. Note of transfer on reverse of Donelson's survey.

Chapter Three: From Bloomery to Blast Furnace

1. R[uth] H. Early, *The Family of Early* . . . (Lynchburg, 1920), 21–24.
2. Ibid., 63–65.
3. Marshall Wingfield, *Pioneer Families of Franklin County, Virginia* (Berryville, Va., 1964), 27.
4. Early, 90–91.
5. U.S. Circuit Court, Virginia District, Ended Cases, 1811, *Trent and Callaway* v. *Dodson's Executors,* VSL.
6. Ibid., Trent's and Callaway, Rocky Ridge, to Dodson, Daltera, and Company, Liverpool, 16 Feb. 1770.
7. Committee of Safety, Ledger, Sept. 1775–May 1776, pp. 42[a], 42[b], Records of the Auditor of Public Accounts, VSL. The committee reimbursed Callaway on 9 February 1776 for money given "Mr. Innes to purchase necessaries for Working the Lead Mine." The exact date of Callaway's appointment as superintendent has not been determined.
8. H. R. McIlwaine, Wilmer L. Hall, George H. Reese, and Sandra Gioia Treadway, eds., *Journals of the Council of the State of Virginia,* 5 vols. (Richmond, 1931–1982), 2:42.
9. Letter, David Ross, Oxford Iron Works, to John Hook, New London, [Oct. 1776], Accession 19904, Personal Papers Collection, VSL.
10. Letter, David Ross, Oxford [Iron Works], to John Hook, New London, 30 Nov. 1776, Accession 19906, Personal Papers Collection, VSL.
11. Charles B. Dew, "David Ross and the Oxford Iron Works: A Study of Industrial Slavery in the Early Nineteenth-Century South," *The William and Mary Quarterly,* 3d ser., 31 (1974): 192–193.
12. Lead Mine Papers, Receipts, 1778, Records of the Auditor of Public Accounts, VSL.
13. Public Foundry, Westham, Ledger, 1776–1779 (1786), pp. 50[a], 50[b], Records of the Revolutionary Government, VSL.
14. Henry County Loose Papers, Henry County Courthouse, Martinsville, Va.
15. *Gazette of the State of South-Carolina,* 25 Aug. 1779, p. 1, col. 1. Courtesy of Frank Horton, Director, Museum of Early Southern Decorative Arts, Winston-Salem, N.C.
16. Bedford County, Order Book 6, 1774–1782, Reel 40, p. 218, VSL.
17. It is not clear whether in these early "lynchings" the accused was hanged.
18. Bedford County, Will Book 1, 1763–1787, Reel 25, pp. 353–356, VSL. The will was written on 29 July 1779 and probated on 27 Sept. 1779.
19. Ibid., 372–375. Inventory and appraisement recorded 28 Aug. 1780.
20. Henry County, Deed Book No. 2, 1780–1784, Reel 1, pp. 140–142, VSL.
21. Henry County, Will Book No. 1, 1777–1779, p. 39, VSL. The will was probated on 23 Nov. 1780.
22. Ibid., 74. Inventory and appraisement recorded 24 Apr. 1783.
23. Henry County, Deed Book No. 2, 1780–1784, pp. 140–142, VSL. Deed recorded 26 July 1781.
24. Frederick Overman, *The Manufacture of Iron* . . . (Philadelphia, 1850), 248.
25. Ibid., 39.
26. Ibid., 44–46.
27. Ibid., 164.

28. Ibid., 164–165.

29. U.S. Census, Industry Schedule, Franklin County, Va., 1820, Reel 228, VSL. No description of the forge in the 1780s has been found; the 1820 account is of a large forge typical of the earlier period. The known annual output of bar iron by the Washington Iron Works forge in the early 1780s is consistent with a forge of the size described here.

30. Henry J. Kauffman, *Early American Ironware: Cast and Wrought* (Rutland, Vt., 1966), 34–35.

31. Arthur Cecil Bining, *Pennsylvania Iron Manufacture in the Eighteenth Century,* Publications of the Pennsylvania Historical Commission, vol. 4 (Harrisburg, 1938), 84.

32. U.S. Census, Industry Schedule, Franklin County, Va., 1820, Reel 228, VSL. The gristmill and sawmill are not mentioned in any known document of the 1780s. They were, however, standard equipment on blast furnace iron plantations.

33. Bining, 170.

34. William Waller Hening, *The Statutes at Large . . .*, 13 vols. (New York, 1809–1823), 9:28.

35. Ibid., 305.

36. Hening, 10:262.

37. Ibid., 397.

38. Thomas Jefferson, *Notes on the State of Virginia,* ed. William Peden (Chapel Hill, 1955), 28. According to Peden's introduction, pp. xiv–xv, Jefferson wrote the first draft of the *Notes* in 1781, then revised and enlarged it until 1784.

39. Franklin and Rockbridge Counties, Legal Records, 1753–1781, Franklin County, *Callaway and Early* v. *John McCoy,* 1786, Accession 6902-a, UVA.

40. Franklin County Suit Papers, *Callaway and Early* v. *John Rentfro,* 1790, VSL.

Chapter Four: James Callaway and the Washington Iron Works

1. Legislative Petitions, Bedford County and Henry County, Records of the General Assembly, VSL.

2. William Waller Hening, *The Statutes at Large . . .*, 13 vols. (New York, 1809–1823), 12:70.

3. Franklin County, Order Book [1], 1786–1789, Reel 27, p. 2a, VSL. The date of the first meeting of the Franklin County Court has been the subject of some dispute, largely because the first two pages of this order book were rebound out of sequence.

4. John Smith Papers, 1724–1923, Accession 2027, UVA.

5. Henry County, Order Book No. 3, Reel 20, 1782–1785, p. 186, VSL.

6. Franklin County, Order Book [1], 1786–1789, Reel 27, p. 10, VSL.

7. Ibid., 117.

8. Ibid., 122.

9. Ibid., 218.

10. Mutual Assurance Society of Virginia, Declarations, Volume 31, Policy 46, Reel 3, VSL.

11. Franklin County, District Court Determined Papers, *Early* v. *Callaway,* 1790, Franklin County Courthouse, Rocky Mount, Va.

12. John Hook Papers, 1737 (1770–1848), 1889, Duke.

13. Land Office, Grants No. 22, 1789–1791, Reel 88, pp. 386–389, VSL.

14. Land Office, Plats and Certificates, 1784, VSL.

15. Land Office, Grants M, 1781–[17]85, Reel 53, pp. 440–442, 445–447, VSL.

16. Bushrod Washington, *Reports of Cases Argued and Determined in the Court of Appeals of Virginia* (Richmond, 1798–1799), 1:39. This is a summary of the case as it was appealed; the records of the High Court of Chancery were destroyed in the Richmond evacuation fire in 1865.

17. Franklin County, District Court Deed Book No. 1, 1789–1802, Reel 17, pp. 74–76, VSL.
18. Franklin County Land Tax Books, 1786, Records of the Auditor of Public Accounts, VSL.
19. Ibid., 1791.
20. Ibid., 1786.
21. Ibid., 1791.
22. Ibid., 1809.
23. James Callaway, "I. Works," to John Smith "at Mr. Callands," Pittsylvania County, 25 Oct. 1787, John Smith Papers, UVA.
24. Legislative Petitions, Franklin County, 23 Nov. 1796, VSL.
25. Samuel Shepherd, ed., *The Statutes at Large of Virginia, . . . 1792–180[7]*, 3 vols. (Richmond, 1835–1836), 2:50–51.
26. Legislative Petitions, Franklin County, 7 Dec. 1797, VSL.
27. Ibid.
28. Ibid., 4 Dec. 1799.
29. Shepherd, 3:43.
30. Legislative Petitions, Franklin County, 16 Dec. 1803, VSL.
31. Undated broadside summarizing the case of *Hook* v. *Hill*, Accession 22291, John Hook Papers, 1772–1829, n.d., VSL.
32. Franklin County, Deed Book 4, 1800–1804, Reel 2, p. 293, VSL. The deed is dated 1801 (no month or day given) and was recorded on 1 Feb. 1802.
33. Ibid., 313–314. Recorded 5 Apr. 1802.
34. Account, Elk Forge with John Hook, 16 Sept. 1796, John Hook Papers, Duke.
35. *Virginia Gazette and General Advertiser*, 26 Dec. 1800, p. 3, col. 5. Courtesy of Frank Horton, Director, Museum of Early Southern Decorative Arts, Winston-Salem, N.C.
36. Legislative Petitions, Bedford County, 9 June 1783, VSL.
37. Ibid.
38. Henry County, Personal Property Tax Books, 1783–1786, and Franklin County, Personal Property Tax Book, 1786, VSL.
39. Henry County Courthouse, 1783 tithable list, Martinsville, Va.
40. Henry County Court Papers, VSL.
41. Ibid.
42. Franklin County, Personal Property Tax Book, 1786, VSL.
43. Ibid., 1801.
44. Ibid., 1809.
45. Franklin County Suit Papers, *Callaway* v. *McGuire*, 1795, VSL.
46. Miscellaneous Collection No. 2, p. 29, Bible Records, VSL.
47. Franklin County, Order Book [1], 1786–1789, Reel 27, p. 88, VSL. Court meeting of 4 Dec. 1786.
48. Franklin County, District Court Determined Papers, *Hairston and Marr* v. *Menefee*, 1790, VSL.
49. Franklin County, Personal Property Tax Books, 1799, VSL.
50. B. Smith, n.p., to Peter Hairston, n.p., 23 Mar. 1786, Wilson and Hairston Family Papers, 1751–1928, Southern Historical Collection, Wilson Memorial Library, University of North Carolina, Chapel Hill.
51. Franklin County Suit Papers, *Callaway* v. *Gillaspey*, 1793, VSL.
52. Ibid., *Callaway and Early* v. *Woods*, 1790.
53. Ibid., *Callaway and Early* v. *Forsythe*, 1787.
54. Bedford County, Will Book 3, 1803–1811, Reel 25, p. 214, VSL.
55. Ibid., 216.
56. Ibid., 228–233.

57. Ibid., 228–229.
58. Ibid., 228.
59. Ibid.
60. Ibid., 232.
61. Ibid., 232–233.
62. Ibid., 231.
63. Ibid.
64. Tench Coxe, *A Statement of the Arts and Manufactures of the United States of America, for the Year 1810* . . . (Philadelphia, 1814), 91.
65. Ibid.
66. Franklin County Suit Papers, *Callaway's Executors* v. *Brockman,* 1812, VSL.
67. Bedford County, Index to Marriage Bonds, 1754–1870, Reel 57, p. 21, VSL.
68. Franklin County Suit Papers, *Callaway and Saunders* v. *Davis,* 1812, VSL.

Chapter Five: Peter Saunders, Ironmaster

1. Marshall Wingfield, *Pioneer Families of Franklin County, Virginia* (Berryville, Va., 1964), 196.
2. Ibid., 196–197.
3. Franklin County, Order Book [1], 1786–1789, Reel 27, p. 2a.
4. Ibid., 27.
5. Ibid., 122.
6. Wingfield, *Pioneer Families,* 197.
7. Date taken from his tombstone in the graveyard at Belle Grove, Pittsylvania County.
8. Wingfield, *Pioneer Families,* 197.
9. Related to the author in April 1975 by the late Miss Ann Saunders Joplin, a descendant of Samuel Saunders.
10. Saunders Family Papers, property of Mr. William Lee, Bleak Hill, Franklin County.
11. Noah Webster, *An American Dictionary of the English Language* . . . (New Haven, 1841), 2:173, definition 11.
12. Louisa Saunders, Belle Grove, Pittsylvania County, to Judge Fleming Saunders, New London, Campbell County, 11 Apr. 1847, Irvine-Saunders-Watts Family Papers, 1745–1914, Accession 38–33, UVA.
13. From her tombstone in the graveyard at Belle Grove, Pittsylvania County.
14. Militia Appointments, 1808, Franklin County, Office of the Governor, VSL.
15. Original commission, Franklin County Records, 1772–1839, Accession 20432, VSL. This commission also appointed as justices William A. Burwell, Daniel Brown, Moses Greer, Jr., Robert Innes, and Robert Woods.
16. Ibid., notation on reverse of commission.
17. Statement of James Callaway, Jr., n.d., Letters Received, Nov.–Dec. 1831; and petition of Lt. George Crump and others, n.d., Militia Appointments, 1808, Franklin County, Office of the Governor, VSL.
18. Statement of Robert Innes, Sept. 1808, Militia Appointments, 1808, Franklin County, VSL.
19. Petition of Anthony Street, n.d., Letters Received, Oct.–Nov. 1830, VSL.
20. Campbell County, Marriage Register No. 1, 1782–1853, Reel 36, p. 143, VSL. Marriage bond dated 18 May 1814.
21. Alice W. Saunders, Franklin County, to Mrs. Mary W. Morris, Campbell County, 26 Sept. 1814, Irvine-Saunders-Watts Family Papers, 1745–1914, UVA.
22. Quarterly Returns of Salt, Lead, Iron, and Cigars, Franklin County, 1815, Records of the Auditor of Public Accounts, VSL.

23. George H. Hull, *Industrial Depressions* . . . (New York, 1911), Appendix Z, Diagram of Production, Consumption, and Prices of Pig Iron in the United States from 1800 to 1911. The price of charcoal no. 1 foundry pig iron sold in Philadelphia (a reliable indicator of the health of the iron industry) reached its highest level since 1800 in February 1815 at $55 per ton. By late summer 1815 the price per ton had dropped to $52.50 and it continued to decline for several years thereafter. The price did not rise again until the end of 1823 when it stood at $35 per ton. The high price of 1815 never again occurred during the antebellum period.

24. Franklin County Land Tax Book, 1815A, Records of the Auditor of Public Accounts, VSL.

25. Franklin County Personal Property Tax Book, 1815A, VSL.

26. Ibid.

27. Articles of agreement, 21 May 1817, Saunders Family Papers, property of Mr. William Lee, Bleak Hill, Franklin County.

28. *Richmond Enquirer,* 31 Aug. 1814, p. 2, col. 5, and *Lynchburg Press,* 27 Oct. 1814, p. 1, col. 1. Courtesy of Frank Horton, Director, Museum of Early Southern Decorative Arts, Winston-Salem, N.C.

29. Franklin County, Deed Book 7, 1813–1817, Reel 4, pp. 268–269, VSL. Deed recorded 6 June 1816.

30. Ibid., 269–271. Deed recorded 6 June 1816.

31. Franklin County, Deed Book 8, 1817–1819, Reel 4, pp. 98–99, VSL. Deed recorded 26 Jan. 1818.

32. Franklin County, Will Book No. 2, 1812–1826, Reel 18, p. 260, VSL. Inventory recorded 1 Feb. 1819.

33. Franklin County, Deed Book 8, 1817–1819, Reel 4, p. 95, VSL. Deed recorded 5 Jan. 1818.

34. Franklin County, Deed Book 15, 1836–1839, Reel 8, pp. 391–399, VSL. Deed recorded 7 Mar. 1837. The Saunders brothers often waited several years to record deeds.

35. Ibid., 400–404. Deed recorded 7 Mar. 1837.

36. U.S. Census, Industry Schedule, Franklin County, Va., 1820, Reel 228, VSL.

37. Frederick Overman, *The Manufacture of Iron in All Its Various Branches* . . . (Philadelphia, 1850), 395–398.

38. Eliot Wigginton, ed., *Foxfire 2* . . . (Garden City, N.Y., 1973), 142–163. The making of a tub wheel is described, drawn, and photographed, and photographs and measured drawings of a tub mill are provided as well.

39. U.S. Census, Industry Schedule, Franklin County, Va., 1820, Reel 228, VSL.

40. U.S. Census, List of Inhabitants, Franklin County, Va., 1820, Reel 14, VSL.

41. Franklin County, Personal Property Tax Book, 1820A, VSL.

42. Almost a hundred pieces of iron from these sprue molds have been removed from Furnace Creek by J. Francis Amos.

43. Collection of J. Francis Amos, Rocky Mount, Va.

44. U.S. Census, Industry Schedule, Botetourt County, Va., 1820, Reel 228, VSL. Reports from furnaces and forges owned by Matthew Harvey, Robert Harvey, Thomas Maybury, and John Tayloe.

45. Quoted in Maud Carter Clement, "The Washington Iron Works," in *Virginia Cavalcade* 11 (Spring 1962): 17.

46. C. H. Turner, Clarksville, to Fleming Saunders, Rocky Mount, 25 Apr. 1832, Saunders Family Papers, 1798–1903, Mss1 Sa878 a 66–165, VHS.

47. Ibid., C. H. Turner, Clarksville, to Fleming Saunders, Rocky Mount, 17 Nov. 1832.

48. Ibid., Caleb H. Turner, Clarksville, to Fleming Saunders, Rocky Mount, 7 Dec. 1832.

49. Ibid.

50. Ibid., Rawlins, Coleman, and Co., Danville, to Fleming Saunders, Rocky Mount, 8 Feb. 1837.

51. Ann Saunders Joplin, "Saunders Iron Furnace," 18 Feb. 1937, p. 2, Franklin County, Iron Furnaces, WPA.

52. Recommendation of Franklin County Court, Letters Received, Oct.–Nov. 1830, VSL.

53. Ibid., Anthony Street to Governor John Floyd, 1 Nov. 1830.

54. Ibid., petition of Anthony Street to the governor and Council of State, n.d.

55. Ibid.

56. Recommendation of Franklin County Court, Letters Received, Nov.–Dec. 1835, VSL.

57. Abraham C. Shelton to Fleming Saunders, 20 May 1836, Saunders Family Papers, property of Mr. William Lee, Bleak Hill, Franklin County.

58. Recommendation of Franklin County Court, Letters Received, Aug.–Nov. 1836, VSL.

59. James Moore Swank, *History of the Manufacture of Iron in All Ages* . . . (Philadelphia, 1892), 189.

Chapter Six: The Iron Plantation

1. S. Sidney Bradford, "The Negro Ironworker in Ante Bellum Virginia," *The Journal of Southern History* 25 (1929): 197–198.

2. Ibid., 199–200.

3. Ibid., 202.

4. Ibid., 203.

5. Franklin County, Personal Property Tax Books, 1822, 1837, Records of the Auditor of Public Accounts, VSL. Only slaves over twelve years of age were taxed.

6. "Saunders Iron Furnace," 18 Feb. 1937, p. 2, and "Old Washington Iron Furnace," 21 Apr. 1936, p. 1, Franklin County, Iron Furnaces, WPA.

7. Ibid., the Farm, 2 Mar. 1938, p. 3, Historic Houses.

8. Ibid., 4.

9. Joseph Martin, *A New and Comprehensive Gazetteer of Virginia* . . . (Charlottesville, 1836), 178.

10. Ibid., 177.

11. Field notebook, George W. Boyd, pp. 84–85, Geological Survey of Virginia, 1836–1841, VSL.

12. Ibid., 87–88.

13. W[illia]m M. Fontaine, "Notes on the Geology and Mineral Resources of the Floyd, Va., Plateau," in *The Virginias* 4 (1883): 178.

14. George H. Hull, *Industrial Depressions* . . . (New York, 1911), 237, "Appendix A (First Part)." All figures and dates are taken from this table, which lists "average prices per gross ton of charcoal pig iron at Philadelphia from 1799 to 1849."

15. Bradford, "The Negro Ironworker," 205.

16. Franklin County, Personal Property Tax Books, 1837–1840, VSL. No records of the two slave transfers have been found in either the Franklin County deed books or in the Saunders family papers. Proof of the transfers lies in the fact that the Hale brothers and Peter Saunders were the only taxpayers to have such large simultaneous fluctuations in the numbers of slaves taxed.

17. Saunders Family Papers, property of Mr. William Lee, Bleak Hill, Franklin County.

18. Franklin County, Deed Book 18, 1842–1844, Reel 9, pp. 91–92, VSL. Deed recorded 20 Sept. 1842.

19. Franklin County, Deed Book 19, 1845–1847, Reel 10, pp. 500–503, VSL. Deed recorded 25 Feb. 1847.

20. Franklin County, Deed Book 18, 1842–1844, Reel 9, pp. 89–90, VSL. Deed recorded 20 Sept. 1842.

21. Franklin County, Personal Property Tax Books, 1841–1842, VSL.

22. Saunders Family Papers, property of Mr. William Lee, Bleak Hill, Franklin County.

23. Ibid.

24. Ibid.

25. Franklin County, Deed Book 19, 1845–1847, Reel 10, pp. 498–499, VSL. Deed recorded 25 Feb. 1847.

26. Pittsylvania County, Deed Book No. 51, 1847–1849, Reel 22, pp. 509–510, VSL. Deed recorded 3 July 1849.

27. Copy of agreement between Peter Saunders and Whitmel P. Tunstall, n.p., 12 Nov. 1846, Saunders Family Papers, 1798–1903, Mss1 Sa878 a 53–55, VHS.

28. Saunders Family Papers, property of Mr. William Lee, Bleak Hill, Franklin County.

29. Louisa Saunders, Belle Grove, Pittsylvania County, to Judge Fleming Saunders, New London, Campbell County, 11 Apr. 1847, Irvine-Saunders-Watts Family Papers, 1745–1914, Accession 38–33, UVA.

30. Date taken from tombstone, Belle Grove cemetery, Pittsylvania County.

31. *Lynchburg Virginian,* 7 June 1847, p. 3.

32. The Farm, 2 Mar. 1938, p. 4, Franklin County, Historic Houses, WPA. Over the years there have been several reports of other hauntings.

Chapter Seven: The Last Years of the Washington Iron Works

1. Date of birth estimated from age given in U.S. Census, List of Inhabitants, Franklin County, Va., 1850, Reel 56, p. 237, VSL.

2. Marshall Wingfield, *Pioneer Families of Franklin County, Virginia* (Berryville, Va., 1964), 198.

3. "Col. Peter Saunders in a/c with Samuel & Fleming Saunders," 20 Jan. 1848, Saunders Family Papers, 1798–1903, Mss1 Sa878 a 39–52, VHS.

4. Franklin County, Deed Book 21, 1849–1851, Reel 11, pp. 70–74, VSL. Deed recorded 31 Aug. 1849.

5. Franklin County, Deed Book 19, 1845–1847, Reel 10, pp. 500–503, VSL. Deed recorded 25 Feb. 1847.

6. Franklin County, Deed Book 21, 1849–1851, Reel 11, pp. 74–76, VSL. Deed recorded 31 Aug. 1849.

7. Peter Saunders, Sr., Rocky Mount, to Fleming Saunders, n.p., 3 Mar. 1848, Irvine-Saunders-Watts Family Papers, 1745–1914, Accession 38–33, UVA.

8. Joseph R. Anderson, Richmond, to Peter Saunders, Rocky Mount, 29 May 1848, Letter Book, Out, 30 Sept. 1847–14 May 1849, Tredegar Company Papers, VSL.

9. U.S. Census, Industry Schedule, Franklin County, Va., 1850, Reel 232, p. 85, VSL.

10. J. P. Lesley, *The Iron Manufacturer's Guide . . .* (New York, 1859), 182.

11. U.S. Census, Industry Schedule, Franklin County, Va., 1850, Reel 232, p. 85, VSL.

12. U.S. Census, List of Inhabitants, Franklin County, Va., 1850, Reel 56, p. 237, VSL.

13. "Saunders Iron Furnace," 2, and "Old Washington Iron Furnace," 1–2, Franklin County, Iron Furnaces, WPA.

14. Ibid., "Saunders Iron Furnace," 2.

15. Fleming Saunders, Flat Creek, Campbell County, to "My dear Boys," Franklin County, 30 Jan. 1852, Saunders Family Papers, property of Mr. William Lee, Bleak Hill, Franklin County.

16. Peter Saunders, Jr., Rocky Mount, to Fleming Saunders, n.p., 3 Mar. 1852, Irvine-Saunders-Watts Family Papers, 1745–1914, UVA.

17. Franklin County, Order Book, 1850–1855, Reel 32, VSL, court meeting of 5 Jan. 1852. The pages of this volume are not numbered.

18. Ibid., 6 Dec. 1852.

19. Franklin County, Will Book No. 8, 1852–1854, Reel 21, pp. 272–277, VSL.

20. Franklin County, Land Tax Books 1856, 1857, Records of the Auditor of Public Accounts, VSL. The tax records reflect the changes made. In 1856 the value of the buildings on the property was appraised at $300, and the total value of the land and buildings was $2,563.02. In 1857 the figures were $2,250 for buildings and $6,437.04 for both land and buildings, because of "improvements."

21. The Farm, 4, Franklin County, Historic Houses, WPA.

22. Maud Carter Clement, "The Washington Iron Works," *Virginia Cavalcade* 11 (Spring 1962): 14; the turn-of-the-century photograph of the house shows a one-story porch.

23. Lesley, *Iron Manufacturer's Guide,* 64. Lesley says the furnace "is to use hematite ore," indicating that the furnace was not yet in blast. Although Lesley's book was published in 1859, he had spent the preceding two years gathering information, so the furnace may have been in operation before the book was printed.

24. *Lynchburg Virginian Tri-Weekly,* 26 May 1858, p. 1.

25. Campbell County, Will Book 12, 1858–1862, Reel 23, p. 78, VSL. Will probated 14 June 1858.

26. U.S. Census, Industry Schedule, Franklin County, Va., 1860, Reel 236, p. 123, VSL.

27. Ibid.

28. Ibid.

29. Franklin County, Order Book, 1860–1865, Reel 33, p. 242, VSL.

30. Date and place of death taken from her tombstone in the Belle Grove cemetery, Pittsylvania County.

Index

Index

The Washington Iron Works
of Franklin County, Virginia was set in Caslon
by Coghill Composition Company, of Richmond, Virginia,
and printed by Eubank Printing Company, of Lynchburg, Virginia,
on 70-pound Mohawk Superfine paper.
The Virginia State Library's logogram *(above)* depicts the stylized
papyrus design of the brass gates at the main entrance to the library.